DATE DUE

OCT 2 9 2009			

Demco, Inc. 38-293

How Women Got Their Curves and Other Just-So Stories

How
Women
Got
Their
Curves
and
Other
Just-So
Stories

Evolutionary Enigmas

DAVID P. BARASH
AND JUDITH EVE LIPTON

Columbia University Press   *New York*

Columbia University Press
Publishers Since 1893
New York   Chichester, West Sussex

Library of Congress Cataloging-in-Publication Data

Barash, David P.
How women got their curves and other just-so stories : evolutionary enigmas /
David P. Barash and Judith Eve Lipton.
p. cm.
Includes bibliographical references and index.
ISBN 978-0-231-14664-7 (cloth : alk. paper) — ISBN 978-0-231-51839-0 (e-book)
1. Women—Evolution. I. Lipton, Judith Eve. II. Title.

GN281.B368   2009
305.4—dc22   2008040139

∞
Columbia University Press books are printed on permanent and durable acid-free paper.
This book is printed on paper with recycled content.
Printed in the United States of America

c 10 9 8 7 6 5 4 3 2 1

To Eva, Ilona, Jacob, Lizzie, Nellie, Sophia, and Yoav

Contents

Preface

Once upon a time, when we were young and the world was so very new and our children were even younger and full of insatiable curiosity, we used to take them on long car trips, such as from Seattle to Lake Louise, and on the road, when we were finished with Raffi songs and Broadway musicals, we would play the Cream of Mushroom Soup Game. It went like this: picture a bowl of cream of mushroom soup (a staple comfort food in our family). It is composed of little beebles of what are supposedly mushrooms in a thick matrix of mush, something resembling a cream sauce. Now, make up a brief discussion of this familiar item from any of a number of perspectives, say, Marxist: How does cream of mushroom soup contribute to the triumph or fall of the proletariat? Is it a bourgeois exploitation of the working class or perhaps the means to a worker's paradise? Or what about a Chicago School of Economics perspective: If cream of mushroom soup flourishes in a free-market economy, does it taste best there, too?

OK, now tell us how cream of mushroom soup can be seen from a modern perspective: What is the role of this soup in history and society? Then a postmodern one: Is it a social construct, devoid of objective reality? What, if anything, is the essence of this soup—the mushroom bits or the mush? How would a postmodern deconstruction of cream of mushroom soup compare with a Buddhist perspective, which holds that the soup is made of nonsoup elements and is empty of intrinsic form? What about a Jewish outlook? No chicken, no soup! A conservative point of view: cream of mushroom soup is a mainstay of any patriotic, God-fearing, Reagan-loving American kitchen. A liberal point of view: cream of mushroom soup is neither local nor organic, and we do not know its carbon footprint. It should probably not be in the kitchen at all.

You can play the Cream of Mushroom Soup Game with anything. How did the leopard get its spots? Why are men and women different? Where did the alphabet come from? How come the armadillo has a smooth "shell," but the rhino is wrinkled? How did the camel get its hump? One can approach these questions from many perspectives, as did Rudyard Kipling, author of *Just-So Stories*, who made up wonderful, imaginative answers to these very things. As we drove along, our children learned to ask hard—and sometimes silly— questions and then make up answers from different paradigms, and they had to learn how these various perspectives worked: Christian, Buddhist, Freudian, Newtonian, Einsteinian, existential, and so on.

We refer the interested reader to Kipling's story "The Crab That Played with the Sea," in which each animal is told to play at being itself: the beaver to play at being a beaver, the cow at being a cow, the elephant at being an elephant, and the turtle at being a turtle. What, then, is the result when people are commanded, encouraged, or—better yet—just plain freed up to play at being human? Our guess is that prominent among the outcomes would be play itself. Dutch historian Johan Huizinga even suggested a Latin name for our species with this idea in mind: *Homo ludens* (man the player). Intellectual play, we concur, is one of the best games of all, and this book is a Play, in Kipling's sense of the word, and an affirmation of Huizinga's hunch.

We can play the Cream of Mushroom Soup Game ad infinitum (and our children might submit ad nauseum), but our favorite paradigm of them all is *evolution by natural selection*, Charles Darwin's very good, very big idea. We are scientists, and the reality is that scientists, despite their public image as humorless, nerdy, and rigidly serious, are among the most playful of people, constantly trying various "mind games" upon the natural world. Through the lens of evolution—or, more formally, the principles of evolutionary biology—scientists have begun to explain how the camel got its hump and how the elephant got its trunk, just as Kipling wrote, but with the added power of science; that is, their search leads to answers that can be tested and either validated (at least provisionally) or proved wrong (falsified). This process is the difference between science and literature or myth: yarn-spinning games can lead to real-world observations or experiments that support or undermine *any* theory. In the end, scientists are materialists who believe that a real, physical universe is run by explicable laws and processes, that living things are the way they are because of how they evolved, and that scientific play is the way we learn how everything works.

In this book, we invite you to join us in an extended game of asking, "Why did women get to be the way they are?" Few things are as scientifically puzzling as the human female, and we'll try to explain why women are somewhat more enigmatic (especially in various aspects of their sexuality) than are men. Then we'll explore how these womanly mysteries may have developed.

This book's narrative arc is simply the human trajectory—we look at women from puberty to old age. The paradigm is evolution—we apply the single, coherent, overriding principle of all biology to the human species. And we'll start the Cream of Mushroom Soup Game: we invite you to partake in some altogether human, wholly scientific play. For each mysterious aspect of human female sexuality, you will find numerous guesses, hunches, suppositions—which is to say, scientific hypotheses—some or all or none of which might be valid. In fact, just when you think we have nailed down the answer, we may surprise you with a "yes, but." We do this on purpose. We want to forestall any premature termination of the game because the bottom-line answer to, for example, the question "How did women get their curves?" is not (yet) apparent.

We also hope to challenge you to join the quest and come up with some answers that are more complete and perhaps more—or less—imaginative than the Just-So Stories you are about to encounter. This is only the beginning of a game or a Play, à la Kipling, not a conclusion.

We thank our children for putting up with our endless games of shifting paradigms. Perhaps it would have been better to play old rock and roll on the cassette player, or maybe we should have taken them to soccer games or Hebrew school rather than to the Canadian Rockies.

We also thank our children for their patience, their game-playing skill, and the various delightful mysteries they still pose, such as when they're going to pass our genes on to the future.

We are indebted to the many biologists, anthropologists, and other scientific players whose ideas have stimulated us and that we herein attempt, however imperfectly, to convey. We are grateful to Marina Petrova at Columbia University Press for responding to our many inquiries and for responding consistently, thoughtfully, and promptly. And we especially thank our editor, Patrick Fitzgerald, for believing in this book and for making numerous suggestions, most of them very good and some of them truly inspired. Nonetheless, we happily take credit for anything that turned out right and want the world to know that whatever didn't, it's all Patrick's fault. We're also deeply grateful to copyeditor Annie Barva, whose eagle-eyed editorial skills covered up a multitude of our sins.

Last but not least, we thank you, dear readers, for joining in this Play.

How Women Got Their Curves and Other Just-So Stories

"A riddle wrapped in a mystery inside an enigma"—this is how Winston Churchill described Russia in 1939. The same can be said today of fully one-half the human race: women. Particularly enigmatic, it turns out, are their bodies.

And so please join us in exploring an array of unsolved evolutionary mysteries, such as: Why do women menstruate? Why do they have breasts when not lactating? Why do they conceal their ovulation, experience orgasm, undergo menopause? These detective stories aren't "who-done-its" but rather "why-is-its," and as it happens, most of them are sexual puzzles as well. Let's be clear: these riddles and enigmas aren't perplexing simply because they involve women, who—as everyone supposedly knows—are somehow mysterious, if only because they aren't men. ("Why can't a woman be more like a man?" wonders clueless Henry Higgins in *My Fair Lady.*) Rather, the traits in question are notable because they are typically found in women and *only* in women—that is, by and large they are biological novelties, not characteristic of other species. They are almost certainly fundamental to being human, stubborn stigmata of the unique evolutionary heritage of *Homo sapiens,* yet neither understood by scientists nor even acknowledged, for the most part, by the public as the puzzles that they are.

Moreover, most people are unaware that the traits in question are biological unknowns simply because nearly everyone takes the most intimate aspects of his or her life for granted, so deeply woven into our substantive human being that they are rarely identified as the perplexities they are. If you were to interview a hypothetical intelligent fish and inquire as to the

I

On Scientific Mysteries and Just-So Stories

nature of its environment, you would probably not hear "It is very wet down here." The evolutionary enigmas of womankind are the ocean in which we swim—and by "we," we do mean everyone, whether male or female.

To be sure, men aren't altogether understood, either. Why, for example, do they develop facial hair (or, alternatively, why don't women)? What is the evolutionary reason for male-pattern baldness? Or for that well-known reluctance to ask directions? Why are erections so predictably unpredictable: often appearing unbidden and unwanted among the young, then disappearing when wanted among the elderly? And does size matter? (If not, why does it seem to be such a big deal, at least psychologically, even if only to men themselves?) These "male mysteries" and many others will have to await another treatment and perhaps the identification of more questions worth asking as well as the unearthing of new leads worth pursuing.

For now, there is little doubt that certain characteristics of women pose enough unanswered biological questions to justify the asking. It is also clear that scientists have come up with enough evidence to reward the process and along the way to tickle the imagination.

Moreover, since each of the enigmas we are about to encounter has more than one possible explanation, but no single one is clearly correct, we'll investigate many possible explanations in every case. Some of these explanations—or hypotheses, as scientists call them—may seem absurd.[1] Others are more reasonable. Some may even turn out to be true.

Most qualify as "Just-So Stories," named after a delightful book of children's tales by Rudyard Kipling, the famous winner of the 1907 Nobel Prize in Literature who was also a militarist, a supporter of British imperialism, and by most accounts a racist, although he did end his best-known poem with "You're a better man than I am, Gungha Din!" The first of Kipling's Just-So Stories, "How the Whale Got His Throat," begins: "In the sea, once upon a time, O my Best Beloved, there was a Whale, and he ate fishes. He ate the starfish and the garfish, and the crab and the dab, and the plaice and the dace, and the skate and his mate, and the mackereel and the pickereel, and the really truly twirly-whirly eel. All the fishes he could find in all the sea he ate with his mouth—so!"[2] The book then proceeds to give fanciful accounts of "how the camel got his hump," "how the rhinoceros got his skin," "how the leopard got his spots," and so forth. Testimony to Kipling's effectiveness as a writer, a Just-So Story came to mean a delightful but empty fairy tale.

Ever since ethologists, geneticists, and ecologists joined together to create sociobiology, sometimes called "evolutionary psychology" when applied to human beings, they have had to contend with the accusation that their work consists of modern-day Just-So Stories, or imaginative accounts of how the biological world came to its current estate, how the various creatures are connected to one another, and—more controversially, at least for some—how the human species fits into this picture. Efforts to understand the intimate details of our own species have especially evoked the skeptical rejoinder "That's merely a Just-So Story" when the hypothesized details are just that: conjectures lacking in supportive data and empirical validation. Among evolutionary biologists, this criticism can be scathing: to call something a Just-So Story is to dismiss it as unscientific poppycock.

We think it is time to stop running from *Just-So Story* as an epithet and to embrace its merits: not that science *ends up* being a Just-So Story, but that it generally *begins* as one, emerging from curiosity, questioning, and uncertainty. In the best cases, it then progresses to reasoned conjecture, to asking "What if?" and "Could it be?" and then, if the imagined story seems worth pursuing and is in fact pursuable, to validation—or, as philosopher Karl Popper and his devotees would have it, to *in*validation if not true—and, if productive, to further refinement.[3] The enterprise is steeped in wonder—a description that includes, not coincidentally, both meanings of this term: an experience of amazement and appreciation ("the wonder of it all") as well as the act of imaginative inquiry ("I wonder if the continents moved" or "I wonder if matter is actually composed of tiny, irreducible particles"). Between wonder (in either sense) and scientific "fact," there are Just-So Stories.

For some, a Just-So Story is an unverifiable and unfalsifiable narrative. As such, it may be great fun, but it is also inherently unscientific. For others—including ourselves—a Just-So Story is simply a story: a tentative, proposed, speculative answer to a question and therefore a clarification of one's thinking, ideally a goad to further thought *and*, not incidentally, a necessary preliminary to obtaining the kind of additional information that helps answer such questions (in the best cases, leading to yet more questions). When this happens—when the narrative is testable and leads to fact-based research—then, in a sense, it is no longer a Just-So Story, but science, pure but rarely simple and more often complicated.

Explanations labeled Just-So Stories are sometimes in fact legitimate empirical questions, which is to say they are falsifiable—if not based on currently available information, then at least potentially so in the future. It bears emphasizing that not all explanations are equally valid and that science arrives at conclusions based on evidence, as opposed to a postmodernist approach in which every "reality" is imagined to be equally valid. However, it sometimes takes a while to determine whether pure speculation, as seductive and appealing as it may be, actually connects to reality. String theory in physics, for example, does not currently have empirical support, and thus, strictly speaking, it may or may not be scientifically valid. But string theory has been immensely productive of additional research even if, according to some of its critics, it may have a downside as well. Such downsides, if and when they occur, are likely associated with "stories" that are scientific dead ends. Phlogiston was just such a dead end, as was "caloric," "ether," and the pre-Copernican, geocentric universe. But the only way to know for certain that a particular path is a dead end is to walk down it a bit and see what happens.

What is the alternative to proposing a Just-So Story, a speculation as to how things came to be the way they are? One possibility, of course, is to say that God did it. Another is to say, like Topsy, that the wings of birds and the echo-location of bats and the eyesight of eagles and the breasts of women "just growed" for no particular reason at all.[4] In this case, we would be stuck with supposed "explanations" that didn't *explain* anything; perhaps, therefore, we should talk about "Just-Growed" Stories.[5] We prefer the Just-So variety, especially when strongly tinged with a whiff of adaptive significance—that is, when the speculation is based on a plausible relationship between the trait in question and how its possessors might have been "positively selected," or how they are likely to have evolved because they enjoyed somewhat more success in projecting their genes into the future. If the speculation is testable, then so much the better: Just-So becomes Just-Right.

Hypotheses based on evolution include predictions about the trait in question: when it is likely to occur, to what degree, the kind of individuals likely to manifest it, and so on. Accordingly, purists (many of them defenders of exactly this approach) may argue that most of the hypotheses considered in the pages ahead aren't really Just-So Stories at all, but rather science. We think they "doth protest too much," that the boundary between Just-So Stories in the Kiplingesque sense and the positing of testable hypotheses in the scientific sense is often indistinct and that Just-So Stories not only often lead to reputable science, but are typically a prerequisite for it.

In this book, we hope to rescue the baby—the thoughtful, imaginative search for explanations—from being seen as contaminated by the supposedly dirty bath water of "unscientific" yarn spinning. We prefer science to "mere guessing," but we'll take the latter any time over the more rigid alternative: keeping silent unless and until the guesses or stories or hypotheses can be fully evaluated and subjected to rigorous statistical testing. Just So you know.

The stories we tell here involve mostly women. Of course, there is nothing new in wondering about them. It has long been claimed that women are mysterious. Freud's description of female psychology as a "dark continent" appears today not only quaint, but patriarchal and patronizing, even though strictly speaking he was correct: women are indeed puzzling when it comes to the workings of their minds. But so are men. At the same time, most people have no idea how many secrets are hidden within women, all of whom pose biological conundrums that are genuinely unique to them not only as individuals, but also *as women.*

In 1999, science journalist Natalie Angier wrote *Woman: An Intimate Geography.*[6] Her excellent book surveys the female body, with different chapters devoted to various anatomic "continents," such as the ovaries, uterus, clitoris, and so forth. Instead of focusing, as Angier does, on what is known about the anatomy of women, we look at what is not known, all the while suggesting possible answers. If Angier's *Woman* is a "geography," examining the corporeal female map, our book explores those regions that are currently off the charts. Referring to such places, ancient cartographers used to write, *Hic Sunt Dracones* (Here There Be Dragons). We say: "Here there be mysteries . . . and here are some possible explanations for them." We thus offer a different kind of map, created by biologists and focused not on female body parts (with one exception), but on dynamic traits such as concealed ovulation, orgasm, menopause, and so on, all united in the personal trajectory of every woman. Our explanations are intended for anyone—male or female—interested in navigating the complexities of human life. We describe the female body as the biological enigma that it is, pointing out the unknown terrain, suggesting where other explorers have gone astray, and proposing new directions.

It has become fashionable—at least in some quarters—to speak of "the death of science," that we have already finished with the big stuff, so all we have left to do is just some "mopping up." We proclaim the opposite: science isn't dead. It isn't even sick, and its mission assuredly is not accomplished. There is a huge amount that we do not know; to a significant extent, in fact, we don't even know what we don't know!

Science courses and even science writing for the lay person often regrettably give the opposite impression because they nearly always present what is known. This approach seems logical, if only because nature does not give up its secrets readily, and scientists are understandably proud of what has been discovered; like our colleagues, we are eager to share the bounty. And yet this approach—so widely expected and dutifully followed—is also misleading because in fact there is much more that we *don't* know. In a perhaps overused simile, science is like a flashlight or a lantern that ideally helps searchers to find their way. But even though the light of current scientific research may be bright, it thus far illuminates only a tiny proportion of the total. We are still largely in the dark.

We mean this neither as a reverential, philosophic statement of epistemic mystery, reflecting the impossibility that science will ever come to grips with the ineffable, nor as a genuflection toward fashionable postmodernist puffery according to which the natural world can be grasped only as a narrow, culture-bound "narrative" because everything—even the basic principles and empirical findings of physics, chemistry, and biology—is "socially constructed." We believe in science and in its capacity to provide yet more illumination—and so do you, if you fly in airplanes, use a computer, or take antibiotics when prescribed. But we also believe that science is most exciting as a process rather than as a recitation of what has already been discovered.

We also believe that it is possible—indeed, essential—for science to operate from a stance of maximum possible objectivity. To understand quanta, molecules, genomes, ecosystems, or galaxies, it is necessary to step outside (metaphorically at least) and treat these subjects as objects. Ditto for the human kidney, brain, and immune system, as well as for the topics to be covered in this book: menstruation, concealed ovulation, breasts, orgasm, and menopause. Accordingly, we wish to plead guilty, right at the start, of objectifying women—or, rather, of objectifying those aspects of women that we are hoping to illuminate and understand—in order to give them and their mysteries the objective, scientific attention they deserve but have only rarely received.

This book is also unusual in that we try to stand in the light while gesturing toward the surrounding dark, making guesses, spinning yarns, telling stories of varying degrees of likely validity. We take this approach not because we think it will always remain dark or because we favor a hushed acceptance or even an embrace of science's limitations. Quite the opposite: we hope to inspire our readers to think about what might be out there just beyond the present circle of science's bright light, possibly even to explore some of the shadows, and (not least) to get some intellectual exercise and have fun in the process.

In our case, that process is biology—evolutionary biology, in particular. Here, too, the nonspecialist reader will encounter something a bit out of the ordinary. Not only will we be examining things that evolutionary science does not know, but we'll be concerned largely with a kind of knowing that may itself be unfamiliar. Thus, when biologists traditionally inquire into the causes of something, they generally mean, "What makes it happen?" For example, many North American birds migrate twice each year: south in winter and north in summer. Trying to understand what makes birds migrate, biologists might investigate changes in food availability, in temperature, and in day length (which, incidentally, is key for most species). Or they might study how the relevant environmental cue influences a bird's behavior: Which senses are involved, which brain regions, which hormones? All these approaches are perfectly reasonable, important, and worthwhile.

But they are also somewhat limited. In particular, they share a focus on immediate causation, involving *proximate mechanisms*—so named because they all deal with nearby, directly causative factors. They tend to neglect another realm of causation, a somewhat more distant but no less valid approach, involving *distal* or *ultimate mechanisms.* Thus, whatever environmental cue hastens bird migration and whatever its hormonal or neural activation pattern, migration would not take place unless there were some underlying advantage—in other words, some evolutionary reason—for doing it. Red-winged blackbirds migrate, black-capped chickadees don't. Yet for both species the days shorten and then lengthen identically; both species have eyes to perceive changes in daily illumination; both have brain regions that can respond to them; both have the same available palette of hormones. The proximate mechanisms that eventuate in migration by red-winged blackbirds clearly do not operate in black-capped chickadees. Why not? What makes it happen in one case and not in another?

Those seeking to answer these questions find themselves in the realm of distal or ultimate mechanisms. In this case, why has evolution equipped red-winged blackbirds with proximate mechanisms that produce migration, but not black-capped chickadees? The difference presumably has something to do with migration being *adaptive* for the former, but not for the latter. In other words, the trait in question—in our example, migration—somehow helps blackbird genes get themselves projected into the future, but would have a very different effect on chickadee genes. Understanding that adaptive "something" is evolutionary biologists' goal, in pursuit of which they would likely study the costs and benefits of migration for the two species. What are the pros and cons of migrating or of staying put if you are a blackbird or a chickadee?

In the pages to come, we are especially concerned with distal or ultimate mechanisms, with what evolutionary biologists call the "adaptive significance" of things—*why* questions rather than *how* questions, not because we don't recognize or value the latter, but simply because as evolutionists, these issues particularly attract our attention and, we trust, will repay yours. Moreover, just as we are contrary enough to look into what science doesn't know as compared to what it does, we find ourselves intrigued by these ultimate questions, which—even in a time when evolutionary science has been wonderfully productive—remain an intellectual road less traveled. This concern with ultimate, distal, evolutionary causation might not make "*all* the difference," but it will make a substantial amount.

Ultimate and proximate causations sometimes become intertwined and, as a result, readily confused. For example, the German public was recently shocked to learn that 30 percent of "their" women are childless—the highest proportion in any country in the world. Moreover, this phenomenon is widespread in the "developed" world, especially in Europe and increasingly in the United States. And it is not a result of infertility: that is, we are talking here about *intentional* childlessness.

Demographers are intrigued. Ardent nationalists are aghast. Most environmentalists (including ourselves) are delighted, as we are at any suggestion that the sky-rocketing human population might someday return, almost literally, to Earth. By contrast, many religious fundamentalists are distressed by the indication that large numbers of women are employing birth control; after all, these very modern women, albeit childless, do not appear at all sexless. And evolutionary biologists (including, once again, ourselves), are asked, "How can this be?" Because reproduction is the fundamental imperative of natural selection, isn't it curious—indeed, counterintuitive—that people are choosing in such large numbers to refrain from projecting their genes into the future when that is precisely the basis for evolutionary success and the linchpin of ultimate biological causation?

The answer is, yes, intentional childlessness is indeed interesting (in fact, fascinating), but it is in no way surprising. Instead of being an evolutionary mystery, it sheds light on what is perhaps the most notable hallmark of the human species: the ability to say "no"—not just "no" to a bad idea, an illegal order, or a wayward pet, but to our own genes.

Intentional childlessness is thus illuminating. When it comes to human behavior, there are actually very few genetic dictates. Our hearts insist on beating, our lungs breathing, our kidneys filtering, and so forth, but these internal, organ functions are hardly "behavior" in a meaningful sense. When it comes to more complex activities, our evolutionary heritage whispers within us; it does not shout orders.

People are inclined to eat when hungry, sleep when tired, and have sex when aroused, but in most cases we remain capable of declining, endowed as we are with that old bugaboo, free will. Moreover, when people indulge their biologically based inclinations, nearly always it is to satisfy an immediate "itch," whose existence is itself an evolved strategy leading to some ultimate, naturally selected payoff. A person doesn't typically eat, for example, with the goal of meeting her metabolic needs, but to satisfy her hunger, which itself is a benevolent evolutionary trick that induces the food deprived to help out their metabolism.

For more than 99.99 percent of their evolutionary history, people haven't had the luxury of deciding on their reproduction: simply engaging in sex took care of that, just as eating took care of nutrition. Evolution, accordingly, endowed *Homo sapiens* with sexual desire, along with food desire, sleep desire, and so forth. But now something new and quite wonderful is on the scene: birth control. Because of it, women can exercise choice at last and, if they wish, save themselves the pain, risk, and inconvenience of repeated pregnancies, or—contra their seeming evolutionary interests—they can indulge themselves rather than their genetic posterity.

This decline in reproductive rates happens again and again, from Nigeria to Nicaragua, when people have the opportunity. "Developed" countries (whose average family size is relatively small) experience even lower birth rates when each child is expected to be outfitted with an iPod and yoga lessons, not to mention a personal trainer. When it comes to our behavior, evolution is clearly influential. But only rarely is it determinative, even when something as deeply biological as reproduction is concerned. Indeed, the trend toward childlessness (or, as many prefer to phrase it, being "child free") is not particularly European, American, or even Western, nor is it strangely "unbiological," but rather profoundly human.

What is nonetheless profoundly counterintuitive is that baby making—doubtless deeply rooted in evolution—should be so readily overridden. But such is the power of modern technology (especially birth control), combined with social mechanisms that convey a discernible payoff to being nonreproductive, all the more so when the proximate pleasures of sex, with or without marriage, remain available.

Although sex looms large in the present book, the traits we discuss are different from the decision to refrain from reproducing; they are much less subject to conscious manipulation and thus in a sense "more biological." Short of drastic surgical intervention, for example, a woman cannot decide to refrain from having breasts. The decision "whether to have children" is, by contrast, rather easy to interpret, and, indeed, the reasons for this decision can be gleaned by simply asking the people involved. In contrast, most women are unlikely to give a cogent answer if asked why they conceal their ovulation or why they experience menopause. This is why we look to science and to evolutionary Just-So Stories in particular to provide some answers.

Finally, note that we examine various evolutionary enigmas not because human beings are fundamentally outside the purview of Darwinian understanding, but for exactly the opposite reason: because evolutionary reality is precisely where women (as well as men and all other living things) reside. When physicists discover something not immediately explicable in terms of their current knowledge, they don't usually give up on mathematics, which provides the framework for their science; rather, they look into the phenomenon more deeply, and using mathematics, they explain what is going on and increase their understanding in the process. By the same token, we have no doubt that the mysteries you are about to encounter will eventually be explained within an evolutionary framework, which has the same gravity within biology as mathematics and, well, gravity have in physics.

But we also agree with Friedrich Nietzsche, who wrote that "[i]n science convictions have no rights of citizenship. Only when they decide to descend to the modesty of hypotheses . . . they may be granted admission . . . though always with the restriction that they remain under . . . supervision, under the police of mistrust."[7] Therefore, as you survey the various hypotheses in this book, treating them—we hope—with intellectual respect, we also invite you to acknowledge their modesty and to view the resulting Just-So Stories with a healthy dose of supervisory mistrust. A story, however entertaining or plausible, is just a story and not a fact until it is supported by hard evidence.

We also recommend another component of healthy mistrust. Beware the temptation to seek *the* explanation for something, especially if that something is as complicated as a trait in a living organism. Even within the seemingly narrow realm of adaptive or "ultimate" evolutionary explanations, causation is often multifactorial. Human evolution, for example, has been accompanied by a substantial increase in brain size from ancient ape to modern *Homo sapiens*. Many explanations have been proposed for this increase: tool use, communication, hunting, early warfare, defense against predators, dealing with the challenges of rapid environmental change or with the complexities of sex and courtship or of sociality more generally. And this list is certainly incomplete. We may never know which of these considerations is the causative culprit.

More to the point, it is very likely that there was no single, solitary, prime-moving cause for the increase in our brain size. Several factors acting together in complex ways probably contributed to the large brains of today's human beings. And the same can be said regarding the causes of women's menstruation, breasts, orgasm, and so forth. Each of these phenomena—like the human brain—may very well have evolved under the influence of several different pressures interacting in various patterns. For the sake of simplicity, in the pages to come we consider numerous hypotheses in turn, one at a time, mostly because the human brain works best that way. But for the sake of reality, keep in mind that the world isn't nearly as linear, unidimensional, or simple as our own species' mental processes, no matter how large those magnificent brains.

This book is the collaboration of a married couple: David P. Barash, an evolutionary biologist and professor of psychology at the University of Washington, and Judith Eve Lipton, a physician and psychiatrist specializing in women's issues. It could not have been written as recently as five years ago. In those five years, findings in biology and in the new science of evolutionary psychology have enabled biologists, psychologists, and physicians to recognize the uniqueness of certain aspects of women's biology compared to other female animals. (Whatever else human beings are, they are animals, too.) Evolutionary thinking has also begun to reveal the hidden sources of this uniqueness, although in nearly all cases it has yielded more questions than answers. But that is precisely what makes science so exciting, and, indeed, one of this book's goals is to identify and share these uncertainties, while keeping a foot—whenever possible—on solid ground.

Just as the surface of our planet is more water than land, the mysteries we are about to discuss are just that, and we are mostly at sea. Accordingly, you will encounter many instances of "maybe," "perhaps," "possibly," "it seems that," and "it appears that," periodically leavened—in spasms of comparative certitude—with "probably" and "it is likely that." Such equivocations do not bespeak any lack of confidence on our part as to the power of evolution by natural selection, which philosopher Daniel Dennett has justly called "the single best idea anyone has ever had."[8] Still less do they indicate any question in our minds that however enigmatic aspects of female sexuality seem, they all are ultimately within the purview of evolutionary science. Rest assured: they will eventually be explained; it's merely a matter of time.

It is said that the Delphic oracle pronounced Socrates the wisest man in Athens—because he knew how much he did not know. By the time you have finished this book, you, too, will be wise beyond measure. Or, at least, if we have done our job, you will be surprised at how much is currently unknown and therefore how much room there is (indeed, how much need there is) for scientifically informed storytelling. Accordingly, let's look at what is known and what is not and hope that there will be an increase in the former, a decrease in the latter, and an intellectually fulfilling journey along the way.

The first mystery begins the moment a young girl starts to become a woman. At some point in adolescence, perhaps as early as age ten or as late as age seventeen, she will begin to menstruate. In all likelihood, she will have been somewhat prepared for this event, informed by parents, teachers, books, television shows, or even pamphlets from her doctor, and also probably misinformed by friends and frightened by her own fear of embarrassment, as in this account from the early teenage hero/heroine of Jeffrey Eugenides's novel *Middlesex:*

2

Why Menstruate?

I was aware that something happened to women every so often, something they didn't like, something men didn't have to put up with. . . . And then one day at Camp Ponshewaing, Rebecca Urbanus climbed up on a chair. . . . We were having a talent show. . . . The sun was still high and her shorts were white. And then suddenly, as she sang (or recited), the back of her white shorts darkened. At first it appeared to be a shadow of the surrounding trees. Some kid's waving hand. But no: while our band of twelve-year-olds sat watching, each of us in camp T-shirt and Indian headband, we saw what Rebecca Urbanus didn't. While her upper half performed her bottom half upstaged her. The stain grew, and it was red. Camp counselors were unsure how to react. Rebecca sang, arms outflung. She revolved on her chair before her theater-in-the-round: us, staring, perplexed and horrified. Certain "advanced" girls understood. Others, like me, thought: knife wound, bear attack. Right then Rebecca Urbanus saw us looking. She looked down herself. And screamed. And fled the stage.[1]

Insofar as an adolescent girl anticipates her first menstruation (technically called *menarche*), she typically does so with a mixture of apprehension and eagerness, concerned—as all children are—about being "normal." Unlike the fictional Rebecca Urbanus, she cannot flee the stage of her own maturation. Nor should she. Despite the tendency, especially in Western societies, primly to ignore menstruation or to consider it somehow embarrassing, a condition best left unmentioned, the reality is that menstrual bleeding is an important part of being a normal, healthy young woman. But even the best-informed adolescent or adult is probably unaware of how curiously "abnormal" menstruation really is when human beings are compared to most other living things. She may be a perfectly typical *Homo sapiens,* but her body's first foray into sexual adulthood takes every woman-to-be into strange, scientifically uncharted waters.

Menstruation isn't unique to human beings, but it is nearly so. Dogs show occasional vaginal spotting during their cycle, but the amount of blood is much smaller, and the mechanism of its release (so-called proestrus bleeding) is entirely different. In fact, it may be that to some extent all mammals shed at least part of their uterine lining between successive ovulations. The Old World monkeys and apes[2] are—not surprisingly—closer to human beings when it comes to menstruation, but for some reason (or reasons) *Homo sapiens* far exceeds any other species in the amount of blood flow, the amount of tissue loss associated with it, and the resulting need to rebuild the uterine lining, known as the *endometrium.* At the beginning of the female reproductive cycle, when a ripe egg is released by the ovary, a complex array of hormones causes the endometrium to grow. If fertilization occurs, the early-stage embryo implants itself deep inside this tissue, and pregnancy begins; if not, then about two weeks after ovulation, the outer layer is sloughed off along with some blood. In nearly all mammal species, either the uterine lining is resorbed and its nutrients retained by the body, or it remains largely available for the next fertilized egg. But not in people.

Admittedly, the blood loss is not huge—about forty milliliters—but it is enough to necessitate additional iron supplements for many menstruating women and is often associated with cramps and other discomfort that most women would just as soon do without. Worse yet, one in ten suffers from endometriosis, an extremely painful and potentially dangerous condition caused by a kind of "backflow bleeding," in which endometrial cells are discharged upward and through the fallopian tubes into the pelvic cavity, where they proliferate each month in response to hormonal changes in the cycle. Given these liabilities and others, it certainly isn't enough to claim, as many older midwives did, that menstruation is "the weeping of a disappointed womb," shedding bloody tears of regret when its yearning for an embryo is thwarted. Moreover, menstruation establishes a time, typically three to five days, during which pregnancy cannot occur. By restricting the available window of reproduction, menstruation thus imposes an additional toll.

An important principle of evolutionary biology is that most traits, if widespread, carry with them sufficient benefits to compensate for their costs. Otherwise, they are liable to be selected against and disappear. Think of the classic scales of justice, replaced here by the scales of natural selection: on one side, the fitness benefits associated with every trait; on the other, its costs. Think, as well, of the biblical "handwriting on the wall"—"You are weighed in the balances, and are found wanting" (Daniel 5:27, American King James version). This is something that natural selection does: it weighs each characteristic of every living thing in terms of whether that trait adds to or subtracts from the success of the species' underlying genes in projecting copies of themselves into the future. After eons of weighing, those traits found wanting are unlikely to be found at all.

Ostensibly neutral characteristics, such as eye color or attached or free-flapping earlobes, are less prone to prompt elimination, but nonetheless ought to deviate increasingly and erratically from the population norm if selection is not there to keep them in line. (After all, the Second Law of Thermodynamics, which states that entropy or disorder increases in natural systems unless energy is available to counteract this process, applies to organisms no less than to nonliving, physical systems.)

Menstruation is notoriously variable not only among women, but even within each individual. Nonetheless, it is a predictable reality for all healthy fertile women. Prolonged amenorrhea, or the absence of menstrual periods, is a sure sign that something is wrong. Let's assume, therefore, that menstruation is neither biologically random and irrelevant nor a manifestation of reproductive failure or the workings of a vengeful deity determined to make women suffer because they are descended from a disobedient Eve.

Here is a quick-and-dirty summary of the human menstrual cycle. It is generally agreed to begin on the first day of vaginal bleeding: Day 1. This *menstrual phase* lasts on average four days, during which the uterine lining, or *endothelium*, is sloughed off with associated bleeding and the loss of small amounts of tissue. Then the *follicular* or *proliferative phase* begins around Day 5. Follicles (literally, any group of cells surrounding a cavity) in the ovary begin multiplying under the influence of several different hormones, notably follicle stimulating hormone (FSH). At the same time, the uterine lining begins to thicken, or proliferate, stimulated largely by estrogens. One or occasionally two follicles become dominant, whereupon the nondominant ones atrophy, and, largely in response to luteinizing hormone (LH), an egg—rarely two—is released: ovulation happens, around Day 14. It marks, in a sense, the midpoint of the cycle.

Then comes the *luteal* or *secretory phase,* which lasts from about Day 15 until Day 28. During this time, the cells that had surrounded and nourished the developing egg become condensed into a corpus luteum ("yellow body") under the influence of pituitary hormones. The corpus luteum then secretes estrogens and, notably, progesterone, which act on that rapidly proliferating lining of the uterus, encouraging its continuing growth and making it potentially receptive to an embryo should one arrive and attempt to implant itself in the lush, secretory tissue.

If no such embryo arrives, the corpus luteum begins to regress, progesterone levels fall, and the uterine lining is shed: menstruation, or Day 1 again, and the dawning of a new menstrual cycle.

Menstruation is clearly the result of a complex and delicate interplay among numerous hormones and target organs. It is not a pathology or a mistake, and it simply cannot have arisen by chance alone. Rather, it must have adaptive significance; in other words, it must have been selected for. Not only that, but it imposes a genuine physiological and anatomical cost; therefore, it presumably carries with it a compensating payoff. But what is that payoff?

Aristotle, in some ways the father of biology as a science, speculated that menstrual blood harbors a substance, the *material prima,* from which an embryo is shaped. Hippocrates argued that men cleanse their blood by sweating, but that women menstruate to remove impurities. Indeed, it was long believed that menstruation served to cleanse a woman's body of "bad humors." From this myth, it was only a small step—via the ancient Greek physician Galen—to the medical (mal)practice of "bleeding" patients to help cure their illness, but which almost certainly hastened their deaths, as happened to George Washington. (Maybe menstruation killed the "father of our country.")

Bleeding, as such, is no longer medically prescribed, but menstruation persists. If, as is now widely believed, menstruation is somehow required to "reconstitute the uterine lining" in preparation for possible implantation during the next ovulatory cycle, why don't other mammals require comparable reconstitution? Why build up an energetically expensive endometrium and then shed it every month?

Is Menstruation an Artifact?

One possibility is that menstruation is simply a peculiar remnant of our primate heritage when our female human ancestors were pregnant more often than not and only rarely experienced a nonreproductive cycle. In this case, menstruation would be more accurately described as vestigial (comparable to the human "tail bone") or simply aberrant than as adaptive. In other words, perhaps primordial hominids simply didn't menstruate very often because most of the time they were either pregnant or nursing. So perhaps menstruation is an artifact of modern birth control, a phenomenon that in "natural" human populations occurred so rarely as to be inconsequential.

The average American woman is potentially reproductive for about 37 years, which equates to approximately 450 menstrual cycles. There is indeed no question that the number is much lower for women in pretechnological societies today, whose experience is doubtless much closer to that of a Pleistocene woman. Hunter-gatherer women today are about 16 years old at menarche, compared to 12.5 years typical among contemporary Americans, almost certainly because of their enhanced nutrition—and, most likely, overfeeding. Not only do contemporary women in technological societies experience menarche earlier, but they also begin childbearing later than their hunter-gatherer counterparts: modern Americans are on average 24 years old when their first child is born, compared with 19.5 years for hunter-gatherers. Women in nontechnological societies nurse for 3 to 4 years, compared with three months (if that) for most Americans; they have a completed family size of 5.9 live births, compared with 1.8 for Americans; and they average 47 years of age at menopause, compared with 50.5 years for Americans. Hunter-gatherer women, then, experience a total of about 160 menstrual cycles in their lifetime, whereas contemporary American women have about 450.[3] Compared to a modern-day American woman, a typical Dogon woman of Mali, Africa, undergoes only about one-third to one-quarter as many menses.[4]

But even 160 menstrual cycles is nothing to sneeze at; multiplied by roughly forty milliliters, they amount to a considerable blood loss. Considering that evolution by natural selection is capable of adjusting such minor adaptive perfections as the number of spots on the wings of a butterfly, it seems most unlikely that even our pretechnological ancestors would have experienced on average more than 100 menstrual events without getting something in return. More precisely, unless there was a fitness benefit to all that menstrual cycling, women whose genetic makeup induced them to refrain from menstruating would have been favored by evolution, and menstruation would long ago have disappeared. Women who menstruated would have been "weighed and found wanting." The fact is, however, that the opposite is true.

Can evolutionary science read the handwriting on the wall?

Is Menstruation a Signal?

There are several possible ways to translate that writing, which is to say that there are several potential fitness payoffs of menstruation. Let's start with the most obvious, but also the least likely. Maybe menstruation is how a woman's body tells her brain that she isn't pregnant. After all, this is one way that menstruation is "used" today, just as, alternatively, it is universally understood that someone whose period is late may well be pregnant. It would clearly be advantageous for a woman to know her reproductive status, and in a world without pregnancy test kits, the presence or absence of menstruation probably gave the first indication.

But this connection doesn't mean that menstruation evolved for that purpose. It would seem terribly inefficient to send such an expensive "all-clear signal" every month rather than utilize simple silence (which would cost nothing) paired with a specific pregnancy indicator when called for. Why couldn't a woman develop a brief case of hiccups or green-tinged earlobes to tell her she is pregnant? She shouldn't have to lose so much of her uterine lining month after month just to alert her to when the bleeding *doesn't* take place! (That would be like being selected to bang our heads against walls so as to get us to pay attention to when we stop doing so.)

It may be noteworthy that although there is no doubt that human beings are the champion menstruators, reproductive biologists are by no means agreed that certain other species don't also do so, albeit less conspicuously. Humans are overt menstruators, as are chimpanzees and some Old World monkeys. Also menstruating, for reasons currently known only unto themselves, are the Transvaal elephant shrew and the black mastiff bat (but you already knew that).

Other species are covert menstruators, sloughing off small quantities of uterine material and blood, the latter often being resorbed in the uterus. It seems only logical for plasma and red blood cells to be recycled in this way rather then being lost forever. Moreover, the fact that human beings are overt menstrual cyclers when presumably they could be thrifty covert recyclers suggests that maybe all that notoriously conspicuous blood isn't so much lost as sent forth to be seen—that is, as a signal.

The signaling hypothesis has three more variants, none of which to our knowledge have previously been suggested. Perhaps monthly bleeding is a sociosexual signal, informing other women that the signaler is entering the realm of sexual maturity and presumably warrants being taken seriously. After all, in many societies the onset of menses is associated with various rites of passage whereby a girl is acknowledged to have become a woman. Perhaps menstrual bleeding is simply a physiological marker of this transition, directed toward other women whose company they are poised to join.

The biggest problem with this hypothesis is that it would seem disadvantageous to the woman sending the signal because female-female competition— although less violent than its male-male counterpart—is nonetheless real, and we might expect such competition to be specifically directed toward others who are just becoming reproductively competent. Thus, it is well documented in the biological literature that among social species, dominant females often suppress the reproduction of subordinates (watch *Meerkat Manor* on Animal Planet if you don't care to read that literature).[5] No comparable phenomenon has thus far been recorded for human beings, although to our knowledge no one has thought to ask whether social dominance (as compared with body mass, fat deposition, and overall health) correlates either positively or negatively with age of onset of menarche.

Insofar as female-female competition might be disadvantageous to a young woman just emerging into reproductive competence, natural selection would presumably have disfavored those who announced—to other women—that they are henceforth available for breeding and therefore are suitable targets to be attacked, intimidated, or otherwise competitively suppressed. Such signaling would therefore as likely be a detriment as an asset to a woman's biological success, especially considering that the supposed signaler is necessarily young and thus particularly vulnerable. One might therefore expect that rather than advertising their sexual maturity, young women would slip into it as inconspicuously as possible.

Maybe menses are indeed a signal, intended not for the general public, but only for those more intimately involved with the signaler—in other words, aimed not at competitor women, but at female relatives who might be able (and, according to evolutionary theory, inclined) to render useful assistance to their younger relative if they have more knowledge of her reproductive status. Thus, menstruation might be a way for a young woman to tell her aunts and cousins that she's ready to find a mate or, on a monthly basis, that she isn't pregnant. We're skeptical, but it would certainly be interesting to find out if menstrual flow is in any way correlated with the presence of potentially helpful—or hurtful—relatives.

In an event, there is no doubt that many societies, perhaps most, have latched onto menstruation and woven taboo-laden cultural traditions around it, and not just for a woman's first "period." Menstruating women are sometimes prohibited from preparing food; not uncommonly, they are segregated into special "women's huts." From the eighth to the eleventh centuries, the Catholic Church didn't allow menstruating women to receive communion. Jewish tradition calls for cleansing in a ritual bath, the *mikva*. It appears that many societies—perhaps most—consider menstruation to be awkward, embarrassing, even shameful, and yet such traditions underscore its reality, in dramatic contrast to the Western style of hiding it. Either way, it remains mysterious how signaling their reproductive condition might confer a benefit upon menstruating women relative to others.

Perhaps the signaling hypothesis might still be saved, however, if the signal were directed toward men, informing them of the same important information. But if so, why not just broadcast the requisite chemical message, possibly via pheromones (or *ectohormones,* molecules specialized to operate between different individuals, as opposed to the better-known *endohormones,* which act inside a body)? Why lose precious blood in the process when sending a subtler and less costly signal would work equally well? It is also true, of course, that insofar as menstruation signals maturation, it does so in a manner that is, if not subtle, perhaps usefully imprecise—indicating fertility without specifying exactly when such fertility is greatest; that is, the precise timing of ovulation remains concealed in human beings (more of this topic in chapter 3). Chimpanzees, however, develop huge genital swellings when they are fertile. They certainly wouldn't need menstruation to drive home the point, and yet they are second only to *Homo sapiens* when it comes to conspicuous menstrual blood flow.

Which leads to another variant on the signaling hypothesis. Maybe the signal is directed toward men, but rather than indicating sexual maturation and availability, it says just the opposite: a loud "no," especially to men seeking sex. Marxist anthropologist Chris Knight advances this idea in a book titled *Blood Relations.*[6] Knight's notion—which, frankly, we find downright silly—is essentially a kind of Lysistrata hypothesis whereby women made themselves (by menstruating) sexually unattractive and unavailable to men, who obligingly responded by going hunting, which they supposedly did while their women were menstruating, and were later rewarded, if successful, by gratifying sex from their women, who of course stopped menstruating by the time the horny hunters returned.

Unfortunately for the Lysistrata hypothesis, there is nothing to support it and many reasons for skepticism. Men aren't especially likely to go hunting when their mates were menstruating, and it isn't at all clear how women who refused sex would have been favored by evolution over those who were more obliging, especially considering that under those circumstances frustrated men would have been selected to look elsewhere for receptive women, who, in turn, would have been genetically rewarded for their receptivity. (A possible answer to this objection is "menstrual synchrony"; see the end of this chapter for its intricacies and enigmas.)

There is, in any event, no reason to think that menstruation exists as a kind of "sex strike," especially because if such were to take place, it should occur in *response* to certain circumstances, such as protein or iron deprivation. To the contrary, menstruation is more likely to *cause* such problems rather than to signal their onset. It seems most unlikely that menstruation is a signal at all.

Of course, there is no question that menstruation, left to its own biological devices, is not hidden, even if, almost certainly, it hasn't been selected as a lurid means whereby women "just say no." It isn't subtle, despite the bizarre efforts made by the modern, Western feminine hygiene industry to provide women with "protection" from their own bodies, to suppress, tip-toe around, or otherwise obscure the natural monthly flow of a healthy woman. At least, it wasn't subtle in pretampon, pre–menstrual pad, pre–"sanitary panties" times, which, after all, constitute 99.99 percent of our evolutionary past. "Nothing could easily be found that is more remarkable than the monthly flux of women," wrote Pliny the Elder several millennia ago. "Contact with it turns new wine sour, crops touched by it become barren, grafts die, seeds in gardens are dried up, the fruits of trees fall off, the bright surface of mirrors in which it is merely reflected is dimmed, the edge of steel and the gleam of ivory are dulled, hives of bees die, even bronze and iron are at once seized by rust, and a horrible smell fills the air."[7]

Such attention, demeaning and inaccurate as it is, lends at least some credibility to the notion that menstruation is so conspicuous that it may still be a signal, albeit one whose purpose—if any—is currently obscure. It also encourages us to look for other explanations.

The Cleansing Hypothesis

One of the most notable suggestions came in the early 1990s from Margie Profet, who theorized that perhaps menstruation is a kind of natural disinfectant, directed at pathogens introduced during sexual intercourse and enabling a woman to "start afresh" every month.[8] Profet herself is an unusual figure in the scientific world, lacking an advanced degree but not at all lacking in either brilliance or chutzpah. (After attempting to solve the riddle of menstruation, she went on to propose that "morning sickness" is an adaptive response of pregnant women whereby they avoid chemicals that might be hurtful to their fetus, and then she won a MacArthur Foundation "genius" award before turning her attention to astronomy.) The title of Profet's attention-grabbing menstruation manuscript in the stately *Quarterly Review of Biology* nicely summarizes what we call the cleansing hypothesis: "Menstruation as a Defense Against Pathogens Transported by Sperm."

It's a nifty idea, not least because it turns around the old canard that menstruation is somehow unclean. According to the cleansing hypothesis, the exact opposite is true: it is the male sexual products—sperm plus semen—along with the physical act of sexual intercourse itself, that is likely to be unclean, with menstruation serving as a hygienic countermeasure.

Profet's notion generated a great deal of attention, in part because it seems to have been the first serious effort to provide an evolutionary explanation for menstruation. In her article, Profet points out that without an appreciation of menstruation's function, it has been all too easy in the past to decry the process as a defect, as did the famous seventeenth-century anatomist Regnier de Graaf (who lent his name to the Graafian follicle, found in the ovary). "The menstrual blood," wrote de Graaf, "escapes by the feeblest parts of the body, in the same way that wine or beer undergoing fermentation escapes by the defective parts of the barrel."[9] Although in less flagrantly demeaning terms, some physicians even today (thankfully a small and diminishing number) advocate this view and consider that menstruation is, if not a symptom of illness, an inefficiency or weakness in women, a defect to be corrected.

The cleansing hypothesis is neither an anti-male screed nor a fastidious critique of sexual intercourse as dirty and dangerous. It begins with a clear-eyed look at the female reproductive tract's vulnerability to infection. After all, the uterus and oviducts are internal organs, richly endowed with delicate tissue and therefore susceptible to being invaded. Moreover, sexual intercourse by its very nature involves the introduction of foreign material deep inside a woman, in the process bypassing many of the body's traditional defenses.

Profet points out that pathogens such as *Chlamydia trachomatis,* which is a common cause of pelvic inflammatory disease and can produce infertility, hitchhike along with other bacteria such as staphylococcus and streptococcus on the tails of sperm. These pathogens and bacteria often inhabit the vagina, where they cause no harm, but if transported to the upper reproductive tract—for example, on an obliging penis—they can be serious troublemakers. Accordingly, Profet proposed that menstruation functions to protect the uterus from infection, and she mobilized an impressive array of facts to support that proposition.

The uterus, by peeling away much of its surface, regularly displaces troublemakers that may be lurking there. Profet points out that the uterus is "designed to bleed," with specialized spiral-shaped arteries and arterioles that constrict—cutting off blood flow to endometrial tissue and causing it to die—and then dilate, promoting copious blood flow that ruptures the tissue and separates it from the underlying uterine wall. The result is not only a mechanism for killing off and then dislodging potentially infected tissue, but also for hosing down and thereby cleansing what remains. (This series of events was first observed in some remarkable research conducted in the late 1930s, in which chunks of endometrium were transplanted from the uteruses of rhesus monkeys into the chambers of their eyes, from which tissue changes were then conveniently observed in conjunction with hormone changes!)[10]

Intrauterine devices (IUDs), a means of birth control, provide further evidence for this hypothesis. They have long been known to cause a substantial increase in menstrual bleeding, which reduced their popularity, although not their effectiveness. Their mode of operation had long been unclear, but IUDs are now known to induce a mild inflammation of the endometrium, which responds with enhanced menstruation, although this effect typically lasts for only a few months after insertion. It is thought that IUDs act as contraceptives in part because they signal or mimic the early stages of implantation. In addition, after an IUD is inserted, an inflammatory response follows, which leads to bleeding and the release of breakdown products from uterine neutrophils (specialized white blood cells), which are toxic to sperm or to implanting embryos.

Inflammation is often an early tissue response to infection and in the case of the body's response to an IUD is consistent with the cleansing hypothesis. Thus, by generating a mild inflammation, IUDs mimic infection, thereby evoking menstruation, and contraception occurs as a by-product. Moreover, if menstrual bleeding is a response to inflammation, then anti-inflammatory drugs should inhibit it, and they do.[11]

It turns out, in addition, that not all folk superstitions about menstruation are in error. Menstrual blood is indeed different from "normal" blood; it lacks many clotting factors, containing, for example, only one-tenth the concentration of platelets (what stops menstrual bleeding is not "normal" coagulation, but the constriction of those handy spiral arteries). Its low coagulability, in turn, suggests that menstrual bleeding happens not just as an unavoidable by-product of a dislodged endometrium, but because the female body possesses adaptations that specifically *promote* bleeding, and not simply as a way of pressure washing the uterus. It turns out that on Day 1, the concentration of leukocytes in menstrual blood is three times greater than that of regular venous blood. These white blood cells, important as part of the normal immune response, are transferred directly to the uterus via menstruation and are thereby made available to do combat with pathogens.

If, as Profet proposes, menstrual bleeding has evolved as a uterine disinfectant, it faces a built-in biological obstacle: many invading pathogens are iron deprived, so surrounding them with blood might give them just the nutrient they require. Interestingly, however, during menstruation the levels of lactoferrin, a chemical that the body normally secretes to help sequester iron and keep it from these microscopic malefactors, are more than twice as high as they are at midcycle.

The cleansing hypothesis makes a persuasive case. If it's a Just-So Story, it's a good one, and, indeed, one can argue that after a hypothesis has been tested to any extent, it has proved its bona fides (whether empirically supported or not) and has moved beyond the *just-so* label.

Just as natural selection weighs fitness costs and benefits, those who seek to assess evolutionary outcomes are obliged to weigh the pros and cons of every suggested hypothesis. Simply because some facts are consistent with a hypothesis doesn't necessarily mean that they provide support for it or "prove" it. (We'll encounter this problem throughout our forthcoming explorations.) As for the cleansing hypothesis, regardless of why menstruation actually evolved, and even if its evolution had absolutely nothing to do with combating uterine infections, it might still have been necessary for menstrual blood to lack clotting factors—or else there wouldn't be any significant bleeding to attract our attention in the first place, or perhaps the endometrial lining wouldn't get dislodged, something that may occur for reasons disconnected from pathogen defense. Similarly, given that pathogens exist, high levels of lactoferrin and leukocytes may well have been biologically necessary as a way of mitigating what might otherwise be pathogen-*promoting* effects of menstruation. If menstruation occurs for some other reason, then much of what the cleansing hypothesis takes as evidence may simply be a series of adaptations against exacerbating infections rather than something specifically evolved to eradicate them.

The cleansing hypothesis leads to a number of predictions. For example, Profet thinks it likely that all mammals menstruate to some degree, with certain ones simply more overt about it than others. More specifically, she proposed that if we look across the animal world, there should be a correlation between a species' mating system and the prominence of menstruation. Species such as chimpanzees and bonobos, which live in multimale groups, should have pronounced menstruation because most females copulate with many different males, thereby increasing their risk of becoming infected by pathogens. In contrast, harem species such as gorillas or socially monogamous ones such as gibbons should be less menstrually dramatic. The first prediction holds; the second, less so. And human beings, the most profligate menstruators of all, should be the most sexually uninhibited, but we aren't. *Homo sapiens* fall between harem and monogamous lifestyles, yet we bleed like bonobos, and even more.

The cleansing hypothesis seems to imply that human beings are more heavily encumbered with sexually transmitted pathogens than are other species or perhaps more vulnerable to them. Or *Homo sapiens* are possibly more inclined than most to engage in sex with multiple partners. (At present, all these questions are open.) Nonetheless, it is widely and accurately said that to have sex with someone is to have sex, as well, with everyone else that one's partner has ever "dated." The more sexual partners you encounter, the greater the probability of acquiring pathogens—hence, the greater the possible need for periodically flushing them away. It wouldn't be strictly necessary for all women to have multiple partners, simply for them to copulate with some men who did so.

Troublesome for the cleansing hypothesis is the fact that certain New World primates such as spider and howler monkeys, which live in multimale groups, hardly menstruate at all. The woolly spider monkey in particular experiences a diversity of sexual partners, rather like chimpanzees. Profet accordingly predicted that this species would menstruate copiously, but it does not appear to do so. Similarly, vampire bats have a multimale mating system, leading to the poetically just prediction that they must not only consume blood, but lose it, too. We have not seen evidence pointing in either direction.

One might also predict that species with concealed ovulation (chapter 3), should have more pronounced menstruation than those with a very limited and precisely identified estrus. Not only might the former have a longer time likely between copulations—and thus less opportunity to introduce possible pathogens—but they would copulate with a smaller number of males. This prediction appears to hold insofar as human beings, whose ovulation is the most concealed of all species, are also the most copious menstruators.

Other predictions come to mind. For example, the normal range of human menstrual blood flow is from ten to eighty milliliters and tends to be consistent for each woman, with some having characteristically heavy periods, others light periods.[12] The cleansing hypothesis would suggest that women toward the eighty-milliliter range would be less likely to develop uterine infections, and vice versa. To our knowledge, no one has yet tested this prediction, but even were it falsified, the cleansing hypothesis might be upheld with the argument that people who are constitutionally more liable to infection are also likely to have heavier menstrual cycles as a result, which therefore reduces what would otherwise be a higher frequency of illness. We might also expect that women in traditional societies, who menstruate so much less often than do modern Americans, would be substantially more subject to uterine infections, assuming they encounter roughly the same number of sexual partners as do modern Americans. There is no evidence that they are.

Here is yet another prediction. Just as Profet suggests that menstrual flow should be heavier in species in which females mate with many different males, one might also anticipate that women who copulate with many different partners should have more copious menstrual flow, either as a general characteristic (a correlation between behavioral styles and menstruation) or varying with each individual's short-term behavior (heavier flow during periods of greater sexual activity and experimentation). We haven't seen any evidence, either pro or con, for such a pattern, although the fact that each woman's menstruation tends to be consistently either "light" or "heavy" suggests that there is little if any correlation between an individual's menstrual pattern and her sexual behavior. It may seem easy enough to survey women and look for a correlation between intensity of flow and variety of sexual partners, but interpreting results would be tricky because if the expected finding is not obtained, one may always argue—and with some justification—that the prediction is too fine-grained and precise to be reasonable.

Oral contraceptives substantially decrease menstrual blood volume, but they do not increase the risk of uterine infection, which argues against the cleansing hypothesis. Then again, "the pill" increases the density of cervical mucus, which may reduce the ability of sperm—and bacteria—to ascend to the uterus. If at this point you are frustrated by the ambidextrousness of these issues, welcome to the club! President Harry Truman once observed that he'd love to meet a one-armed economist because whenever he asked his advisers for the likely consequence of a particular domestic policy, each response would be followed by "But, on the other hand . . ." Unlike Mr. Truman, you don't have to commit yourself to a particular point of view, and we hope that instead of experiencing Trumanesque annoyance, you will revel in the many-handed-ness of women's evolutionary enigmas and even in the pluralistic plausibility of the explanations.

Finally, the possibility that menstruation does in fact contribute to uterine cleansing offers some intriguing implications for women's health and medical practice. Profet points out, for example, that "the uterus appears to be de-signed to increase its bleeding if it detects infection: Human uteri that become infected (or otherwise inflamed) bleed more profusely, bleed on more days per cycle, and often bleed intermittently throughout the cycle." Bleeding at the wrong time nearly always indicates that something is wrong, but is bleeding part of this "wrongness" or part of the body's response to it? Assuming the for-mer, according to Profet, would be "blaming firemen for a fire."[13]

This distinction is not merely academic because physicians must decide whether menorrhagia (unusually heavy menstrual bleeding) and meteorrha-gia (intracyclic menstrual bleeding) should be artificially treated and curtailed, as by adding clotting factors, or allowed to continue. Dysmenorrhea (painful menstruation) is caused by uterine contractions that are unusually strong, not by bleeding per se. Menorrhagia and meteorrhagia are the major symptoms of endometritis, inflammation of the endometrium. This potentially serious con-dition most commonly results from infection of the oviducts with a pathogen such as *Chlamydia.*

If Profet is correct that uterine bleeding is an adaptive response to infection, then by interfering with menstrual bleeding, modern medicine may be un-dermining a useful body function. Evolutionary logic has pointed out similar dilemmas—for example, what to do about fever. The traditional, nonevolu-tionary view is that raised body temperature is a symptom of inflammation or infection and therefore something that should be counteracted. In at least some cases, however, fever is part of the body's defense against invaders, as shown by the fact that certain pathogens have a harder time dealing with raised temperature. Hence, taking aspirin to fight fever can actually get in the way of the body's efforts to fight disease. Turning things around, however, physicians might want to consider that enhanced bleeding associated with uterine infection, may result from a woman's menstrual machinery having been hijacked by a pathogen that needs iron or some other blood-borne nutrient, in which case it would be medically indicated to inhibit such bleeding.

The cleansing hypothesis also implies that sexually active women with amenorrhea (e.g., serious athletes) are especially at risk of uterine infection and may therefore want to be particularly alert to its symptoms and to use barrier protection. What about sexual intercourse when women are lactating, already pregnant, or postmenopausal and therefore are no longer receiving the presumed benefits of menstrual cleansing? Are they more at risk of uterine infections? The cleansing hypothesis would suggest that they are.

And what about doing away with menstruation altogether? The cleansing hypothesis argues strongly against this and would thus urge caution about the new birth-control pill Lybrel, the first oral contraceptive designed to be taken every day, with no pill-free intervals. Part of Lybrel's appeal to many women is that it prevents regular periods. Insofar as modern women in technological societies experience many more menstrual cycles than do women in traditional cultures and are therefore supposed to have moved farther from the "natural" human experience, maybe some menstrual suppression would be worthwhile. But inhibiting menstruation in otherwise healthy, nonreproducing women is uncharted territory. If the pills fail, a woman might not know if she were pregnant (see the signaling hypothesis), and even if they work as intended, the cleansing hypothesis predicts that users might be more subject to uterine infections.

A Matter of Efficiency?

For all its commonsense appeal, the cleansing hypothesis is controversial and far from conclusive. Three years after it was published, anthropologist Beverly Strassmann pointed out a number of seeming flaws wherein logical predictions that follow from the cleansing hypothesis are not supported by the available evidence.[14]

Strassmann noted, for example, that if menstruation serves to combat pathogens, whether by flushing them away or zapping them with antimicrobial agents, there should be fewer "bad bugs" present after menses than before. In fact, however, the data point the other way: if anything, fewer bacterial species are present in the uterus *before* the supposed cleansing action of menses. However, number of species isn't the same as the actual number of bacteria. The former is relatively easy to determine, whereas the latter may be more important when it comes to causing disease. Strassmann also emphasized that blood is an excellent culture medium for bacteria, providing not only iron, but also sugars, proteins, and amino acids useful for microbial growth.

Toxic shock syndrome, for example, is especially associated with the bacterium *Staphylococcus aureus,* which thrives on the nutrients in blood and—tragically—in menstruating women, who are the ones subject to this devastating disease. But then again, tampons are a very recent development, not part of our evolutionary past, and they change the intravaginal microbial ecology, typically in troublesome ways. In pre–feminine hygiene societies, menstrual blood and its nutritive load of iron would pass safely out of the body rather than accumulate on tampons.

Strassmann has studied the Dogon of Mali, among whom sexual abstinence is practiced for only a month or two after childbirth, although lactational amenorrhea (ovulation inhibited by nursing) lasts an average of twenty months. As a result, the Dogon people—along with many others who lactate extensively, menstruate infrequently, yet still engage in sexual intercourse—should, according to the cleansing hypothesis, have a permanent epidemic of uterine infections. They don't.

Let's face it, the uterus is in a difficult situation: it must protect itself against invading pathogens while not mounting such a robust defense that it subjects sperm to withering friendly fire. As part of the female reproductive tract's balancing act, the cervix produces thick, acidic mucous that prevents bacteria from ascending to the uterus—a defense system that needs to let down its guard during sexual receptivity. When it comes to uterine infections, a woman's cervix is a major protective structure, preventing microorganisms from gaining access to the upper reproductive organs. Strassmann points out that "menstrual bleeding dissipates the cervical mucus—which makes it easier, not harder, for sperm-borne pathogens to ascend to the uterus."[15] Another strike against menstruation as an anti-infection tactic.

And yet does this point argue against the cleansing hypothesis or emphasize the role of competing pressures?

Strassmann agreed with Profet that the copiousness of menstruation should increase with the promiscuity of the breeding system. After conducting a detailed survey of twenty-five different primate genera, though, she found no correlation between degree of promiscuity and intensity of menstruation. Macaques,

EVOLUTIONARY HYPOTHESES EXPLAINING WOMEN'S MENSTRUATION

Vestigial: an artifact of modernity
Signaling 1: absence of pregnancy
Signaling 2: to other women
Signaling 3: to relatives
Signaling 4: to men (maturation)
Signaling 5: to men (Lysistrata)
Cleansing: combating pathogens
Metabolic efficiency: lowering costs
One-way street: terminal differentiation
Competence test: evaluating the embryo
Synchrony 1: predator overloading
Synchrony 2: encouraging monogamy
Synchrony 3: female-female competition
Synchrony 4: social bonding
Synchrony 5: social facilitation
Synchrony 6: synchrony doesn't exist

baboons, and chimpanzees fit the prediction, but many other primates—especially the prosimians (such as lemurs) and New World monkeys—do not. It doesn't seem that copious menstruation has evolved along with high promiscuity, although promiscuity, in turn, should lead to more potential infections.

Strassmann also noted that the vagina and cervix do not bleed even though they have a higher pathogen load than the uterus. In addition, only mammals menstruate. Why not insects, reptiles, birds—all of which experience internal fertilization and presumably are susceptible to infection? Accordingly, maybe menstruation isn't a pathogen defense, but a consequence of the fact that the endometrium becomes highly vascularized in preparation for implantation. Supporting this view, Strassmann pointed to the Transvaal elephant shrew, a small, strange mammal that clearly menstruates, but in which bleeding occurs only from the small part of the uterus at which implantation actually occurs. If such bleeding were an antipathogen adaptation, it should occur throughout the uterus—that is, unless for some other reason suitable arterioles are found only in this area.

Elephant shrew (*Elephantulus rupestris*), another menstruating mammal. Photograph by Darrel Plowes.

With the cleansing hypothesis looking shaky, Strassmann then proposed her own idea, which we dub the metabolic efficiency hypothesis. The argument is simple and logical, even if not very exciting: "The uterine endometrium is shed/resorbed whenever implantation fails because cyclical regression and renewal is energetically less costly than maintaining the endometrium in the metabolically active state required for implantation."[16] In short, it is energetically cheaper to dump the uterine lining and grow a new one each month than to keep the old one going. Evidence in support of this view includes the fact that oxygen consumption in a postmenstrual uterus is only about 14 percent that of a fully prepared endometrium. In addition, a woman's overall metabolic rate is about 7 percent lower during the follicular (preovulatory) phase when the uterine lining is regressed than during the luteal or secretory phase when it is actively growing. According to the metabolic efficiency hypothesis, then, a woman's metabolic economy alternately revs up and down, thereby economizing on the costs of being prepared to reproduce. Strassmann estimated that twelve months without menstrual cycling saves energy equivalent to a woman's calorie requirement for nearly a half-month. (The oft-noted tendency of postmenopausal women to gain weight may even occur because they are no longer wasting calories constructing a costly endometrium and then tearing it down again.)

Energy is the basic currency of life, and there is abundant precedent for living things to do whatever it takes to conserve it. Consider the Burmese python, which goes a very long time, sometimes months, between meals. During such dry spells, in an energy-saving adaptation, these animals literally stop maintaining a functional intestinal lining. Similarly, the ovaries and testes in many birds regress during the nonbreeding season to save the energy of maintaining them and to reduce the birds' weight. The human uterus is in a uniquely favored position compared with, say, the lungs or kidneys: like the intestines of a fasting Burmese python or the gonads of a seasonally breeding bird, it doesn't need to function on a regular basis to keep our bodies going. Moreover, if the metabolic efficiency hypothesis is correct, the highly vascularized uterine lining demands so much energy that its maintenance becomes a liability. Maybe people are pythons at heart—or rather, at uterus.

It is also possible that metabolic efficiency itself is a by-product of menstruation's having been selected for some other reason. Thus, assume for a moment that the cleansing hypothesis is correct after all. Evolution would then generate the process of uterine shedding and bleeding, which would produce metabolic efficiency as an incidental consequence, not as the naturally selected target. By the same token, the alternate contractions and relaxations of the human heart have obviously been selected to pump blood, not to produce heartbeat sounds. The heart's characteristic "lub-dub" is merely an incidental and presumably unavoidable by-product.

Other hypotheses for menstruation have been suggested. Perhaps it evolved as a way of ridding the body of incipient cancers. Cancer is especially likely in rapidly dividing cells, including, perhaps, those in the uterine lining. After all, these cells are stimulated to reproduce rapidly under the influence of estrogen. But there wouldn't be any such rapid cell proliferation if the endometrium weren't sloughed off so frequently in the first place, which, in turn, makes it problematic to argue that regular removal of the endometrium is adaptive as an anticancer tactic, considering that the endometrium wouldn't be vulnerable to cancer if it weren't being regularly removed and then so vigorously regrown.

Scientists are supposed to follow the facts and not to have favorite hypotheses. We try to do just that, but we confess that certain ideas are more appealing than others, and for us part of that appeal derives from our evolutionary bias. Of course, the metabolic efficiency hypothesis is bona fide, 100 percent evolutionary, too, in that it posits that menstruation occurs because it provides for an adaptive conservation of metabolic energy. (Insofar as doing so resulted in metabolic savings that outweighed the costs, women who regularly sloughed off and then regrew their uterine linings would have been more fit and therefore would have left more descendants than did others who refrained from menstruation.) But this hypothesis identifies menstrual bleeding as essentially a by-product of natural selection for something else—efficiency—and as such, it simply isn't as attractive (to us) as the signaling or cleansing hypotheses, which place the phenomenon itself front and center as the apple of evolution's eye.

So we grudgingly grant that Strassmann is correct that Profet was something less than prophetic (although we give her an "A" for effort). Although Strassmann also seems correct in pointing out that it is more costly to maintain a rich uterine lining than to create a new one for the next reproductive go-round, our guess is that her metabolic efficiency hypothesis does not completely explain menstruation and maybe doesn't even partially explain it. It implicitly assumes that there's no way for a prepared endometrium to tamp down its activity and remain quiescent until the next early embryo comes along. But why must the human uterine lining go from riches to rags?

Even if a tear down/build up cycle is somehow unavoidable in order to provide a suitable implantation site while avoiding the costs of keeping the site in tip-top, ready-to-go condition, why not stop cycling when there are no men in one's life? It wouldn't seem beyond the ingenuity of natural selection for our species to have evolved, for example, an endometrium-building response to male pheromones or to the physical stimulation of intercourse itself. Why can't the uterine lining stay quiescent between ovulations or be constructed on an as-needed basis, going into action immediately upon implantation, instead of relying on a living welcome mat that is so expensive to retain that it is cheaper to destroy and rebuild it every month? And besides, why do human beings menstruate so copiously compared to other mammals? Is there something special about our species that requires such an expensive endometrium?

Another question also suggests itself at this point: Why wait so long before the next ovulation? Every female infant is born with her lifetime complement of eggs, some of which eventually mature and are released by an ovary, one at a time, about every twenty-eight days at the midpoint of each cycle. The metabolic efficiency hypothesis posits that it is too energetically expensive to maintain the uterine lining while waiting for the next egg and possible embryo. But why is that interim so long? The energetic investment in maturing an egg is trivial, and it isn't obvious why one can't be made available every week or even every day—at least for a sexually active woman—until pregnancy occurs. And if the latter were the case, wouldn't it avoid the metabolic hassle of constructing a new endometrium? Lots of other body tissues, such as muscle and breast, have no difficulty regressing when not needed, without being lost altogether. If saving energy is so important, then why not suspend the metabolic activity of the uterine lining instead of giving up on it altogether and expelling it?

Part of menstruation's enigma and the need for all these hypotheses is that it poses a kind of Sisyphean dilemma. According to Greek mythology, Sisyphus was condemned to spend eternity pushing a heavy rock up a steep hill, only to have it roll back down each time. A menstruating woman finds herself constructing a snazzy, energetically expensive endometrium each month, only to have it dislodged again and again. Sisyphus had no choice, and neither do most women. But evolution presumably did.

A One-Way Street?

Looking more deeply into the events surrounding menstruation, we find that immediately after bleeding stops, the uterine lining begins to regrow (entering its proliferative phase) largely under the influence of estradiol, the major form of estrogen in humans, which is produced by the ovaries. By the time of ovulation, spurred on by a pulse of LH from the hypothalamus (in the brain) and then supported by progesterone from the ovaries, the endometrium begins the secretory phase, producing nutrients that a newly arrived embryo will need until it gets itself established and begins to obtain food from the mother's blood. At least one biologist—Colin Finn, trained as a veterinarian and specializing in animal reproduction—has argued that when the uterine lining, at the peak of its secretory development, is primed to accept a fertilized egg, roughly two weeks after ovulation, it is irretrievably committed to its job.[17]

At this point, the story goes, if no such passenger arrives, the endometrium is just too far gone down the reproductive road to wait or even to slow down. It simply has to rush along, get sloughed off, and then be replaced as the next cycle begins. But claiming that the human endometrium is just too bulked up to be reabsorbed or too differentiated to be reused and that menstruation has to happen in consequence is equivalent to saying that menstruation has to happen because it has to happen. Why have human beings evolved to produce a too hefty or too differentiated uterine lining in the first place?

Maybe the reason for heavy menstruation is all in our heads—literally. That is, our large developing brains require lots of oxygen, which in turn necessitates an especially thick, richly developed endometrium, whose abundant blood vessels are needed to transfer all that oxygen, but which unavoidably gets so big and so thick that it simply cannot be reabsorbed. This connection may explain—or at least, be consistent with—the fact that the great apes, which have the largest brains, also do the most menstruating. It also agrees with Finn's view that the invasiveness of implantation in our own species requires loss of the endometrium because the uterine lining must become "terminally differentiated" to prepare itself for implantation.

In any event, it may well be significant that among human beings implantation occurs in the deep layer of the uterine lining, the stroma, rather than more superficially in the epithelium, as in most mammals.[18] In many other species that don't menstruate, however, the endometrial lining doesn't differentiate irreversibly until *after* implantation, if ever, and even mammals that lack invasive implantation undergo growth and regression of the endometrium (perhaps to save metabolic energy?). Furthermore, only chimps and people undergo such dramatic differentiation of the uterine lining, technically called *decidualization*. Other species that experience a kind of menstruation, such as rhesus monkeys and baboons, do so without their endometria achieving this supposedly irreversible state. It therefore seems unlikely that human menstruation is necessitated by a wayward or headstrong uterus that has somehow gone too far down a one-way street.

A Competence Test?

Time for a final hypothesis. It is our favorite, in part because we came up with it.[19] Under this view, menstruation is a competence test: not of the woman, but of any fertilized egg she might be carrying.

Let's go back to the fact that menstruation is brought about by the natural decline in progesterone beginning a week or so after ovulation and ask, "What keeps menstruation from happening?" Answer: pregnancy. (Other things do so, too, such as illness, high stress, or intense levels of exercise, all of which exert their effect by inhibiting ovulation. Amenorrhea—lack of menstruation—doesn't cause anovulation, lack of ovulation. Nearly always, it's the other way around.) And how does pregnancy do the trick? If a fertilized egg succeeds in implanting itself in the uterine lining, it immediately begins producing a special hormone as though its life depends on it—because it does.

The newly implanted embryo pumps out a hormone unique to human beings, human chorionic gonadotropin (HCG): *chorionic* because it is made by the chorion, the part of the placenta that is produced by the embryo, and *gonadotropin* because it acts upon the gonads. This hormone is chemically very similar to LH produced in the hypothalamus, which earlier in the cycle kick-started ovulation and then kept the ovaries producing progesterone. Progesterone underwrites the endothelium's embryo-friendly secretory phase, and the decline in progesterone triggers menstruation. This is why HCG and its resemblance to LH is so important to the embryo: in a fertile cycle, HCG picks up where LH dropped off. It keeps the ovaries making progesterone, which in turn keeps the uterine lining in place, which in turn provides a home for the embryo. HCG—produced by the embryo, not by the mother—is even more potent, molecule for molecule, than LH. It is also much more resistant to being degraded and actually results in an *increase* in progesterone output from the ovaries. Moreover, HCG is produced in such quantity and so reliably that it is the basis for home pregnancy tests. Also worth noting: the birth-control drug RU 486 prevents pregnancy by blocking progesterone receptors in the uterus so that menstruation occurs even in the presence of HCG.[20]

This is where things get interesting for the competence test hypothesis and challenging for the embryo being tested. HCG is a large protein molecule (with a molecular weight of around 39,000), too big to pass into a woman's body by crossing her cell membranes. It must be secreted directly into her blood, which is why the embryo can't guarantee its own reception in a welcoming uterus simply by secreting HCG immediately after fertilization or while traveling down the fallopian tubes. It has to wait until it arrives at the endometrium and digs itself in. "Digging in" is quite a challenge, and that may be just the point.

Implantation separates the mice from the men, the wannabes from the pregnancies. In fact, it really does separate mice—and nearly all other animals—from human beings because it is much more demanding in *Homo sapiens* than in other mammals. Among most nonhuman mammals, the early embryo develops within the uterus, making relatively superficial contact with its outer lining, whereas in human beings genuine implantation takes place in the endometrium, which requires actual deep penetration into the inner uterine layers. Because this process is so invasive, it is much more difficult, involving a delicate dance between maternal tissues and embryonic capability.

The former must be receptive, but at the same time (literally) the latter must do the work.

If the embryo successfully implants itself in the uterine lining, it immerses itself directly in pools of maternal blood and starts getting the nutrients it needs. Equally important for its survival, the embryo—technically called a *trophoblast*—gives as well as takes: it dumps HCG into the mother's circulatory system, which, as described, keeps the system going. When it comes to implantation, *Homo sapiens* are at one extreme of the mammalian spectrum. Our species has evolved so that if the embryo fails to dig deep and well and is therefore unable to inject HCG from itself to the mother, it gets washed away in a flood of menstruation. Rather than "hold on tight, here comes a tidal wave," it is more like "dig in deep and prevent the flood." Instead of "sink or swim," think "implant or get washed away," not because implantation is a way of holding on, but rather because successful implantation is a way of preventing the menstrual flood in the first place. According to the competence test hypothesis, evolution has favored menstruation as a way of eliminating any embryos that cannot pass this test.

In health classes, students learn that menstruation happens when pregnancy doesn't. But maybe menstruation takes place so that pregnancy won't if the embryo isn't up to snuff.

It's not easy to assess this idea or even to propose what sort of evidence would confirm or refute it. One possibility is to look into success rates, the prediction being that although many eggs are fertilized, few are implanted. This appears to be the case, although the data are very limited. One study found that 62 out of 198 pregnancies (31 percent) diagnosed by the presence of HCG failed to come to term. The researchers further concluded that one-fifth of embryos beginning implantation fail to make it.[21] Clearly, implantation is a major speed bump on an embryo's journey to gestation. Prior to implantation, the risks may be even greater. A now-classic study was memorably titled "Thirty-four Fertilized Human Ova, Good, Bad, and Indifferent, Recovered from 210 Women of Known Fertility."[22] This research is now fifty years old, but it remains one of the very few to report on "biological wastage in early human pregnancy" and to do so with a reasonably large sample. It concluded that nearly one-half of fertilized eggs fail to implant. An Australian study using somewhat different methodology concluded that the rate of failure may be even higher: of eighteen early-detected fertilizations, only four resulted in pregnancy.[23]

It is extremely difficult to estimate the percentage of early embryos that are lost, especially if the question is "How many are lost before HCG is produced?"

because it is the existence of HCG that is used to pinpoint the presence of an early embryo in the first place. But the proportion that fail to make it through implantation is certainly high, quite likely in the range of 50 percent. Some of these invisible, lost battalions may have perished just because their timing was off and they weren't perfectly synchronized with their uterine targets, but the likelihood is that most of them were either genetically defective or otherwise developmentally aberrant, unable to complete their Big Dig.

Just as researchers have a very hard time assessing the existence of a just-fertilized egg, it makes sense that there is no way prior to implantation for a woman to assess accurately the competence of a just-fertilized egg; after all, it is floating about, unattached. Besides, even the most messed-up preimplantation embryo isn't making any metabolic demands on the mother. Implantation offers the first good opportunity for it to do so.

If so many early embryos are incompetent, why hasn't natural selection favored releasing multiple eggs simultaneously so that among the many eager implanters at least one will be successful? Probably for the same reason that one of the problems of "assisted fertility" procedures is that women experience significantly higher risk of twins, triplets, quadruplets, and so on. Not coincidentally, it is standard procedure in fertility clinics to introduce several fertilized eggs precisely because so many fail to implant. The reality is that human beings, like our other great ape cousins, aren't well adapted to multiple births, and as with in vitro fertilization, multiple simultaneous in vivo ovulations might well produce multiple pregnancies.

It also makes sense that human beings should be especially prone to testing their offspring for competence and to do so early on. Implantation is a kind of Rubicon, which, once crossed, commits the mother to continuing investment, and among *Homo sapiens* that investment is immense. Nine months is a long time to carry and nourish a fetus (although some species have gestation periods that are even longer). In addition, birth is hardly the end of a human parent's responsibility: measured in sheer caloric cost, lactation is several times more expensive than pregnancy, not to mention the looming later costs of diapers, orthodonture, and college! If we stick more seriously with basic biology, the reality is that human beings are unusual—indeed, unique—in the amount of "parental investment" expended on each child, which may in turn help explain why we are also unusual—albeit not quite unique—in our penchant for biparental care and for expending such care not just over years, but over decades.

Nonetheless, it may seem counterintuitive that selection would favor prospective mothers' making it difficult for their potential progeny to get a toehold.

One of the more important evolutionary insights in recent decades, however, has been that parent-offspring conflict is every bit as predictable as is parent-offspring coordination. Thus, in a now classic article, evolutionary theorist Robert Trivers took the commonsense view that parents and offspring are necessarily on the same side and turned it around, showing that they are also necessarily inclined to disagree.[24] It's a matter of seeing the parent-offspring glass as both half full and half empty. The half-full component is due to the fact that each gene within a parent has precisely a 50 percent chance of being present within his or her child. At the biological level, this fact is why parents care for their children, and, indeed, from an evolutionary perspective, it is "why" they produce them in the first place: genes that do not project themselves into the future via children would generally face a rather dim future. At the same time, however, there lurks the genetic half-empty glass, the 50 percent nonrelatedness.

If we were to reproduce asexually so that our offspring were genetic carbon copies of ourselves, there would be little biological basis for conflict, just as there is deep agreement between our liver and kidney cells, all of which are the products of asexual reproduction by a single fertilized egg. But as Trivers points out, a developing embryo is only 50 percent related to its mother, but 100 percent related to itself! As a result, the embryo would be selected to devalue maternal costs by a factor of two compared to its own self-regard—in other words, it would be inclined to demand twice the resources that the mother would be selected to provide. For their part, mothers (and fathers, too) would certainly be nudged by natural selection to contribute to their offspring, but not without limit. They would be more "fit," in the evolutionary sense of having more long-term genetic success, if they partitioned their investment in offspring so as to avoid spending too much on any one if doing so diminished their ability to care for others and for themselves.

The result is a tug-of-war, with offspring selected to garner more parental care than parents are inclined to provide, and parents selected to be parental, but not to go overboard about it. "Unconditional love" is a delicious concept, but we are not at all convinced that it has ever existed or that it should; numerous situations and actions are liable to diminish love, even between the most loving. In this sense, *all* love is conditional. The very earliest parent-offspring interactions, between a mother and her not-yet-implanted embryo, fit perfectly into biological theory that love (which, after all, is another way of describing the inclination to affiliate with another) is conditioned upon the embryo's proving that it is literally worth being loved. It would not be adaptive to throw good love after a bad trophoblast.

Once a child is born and emotional connections made, it is difficult and perhaps even immoral to cut one's losses. But before implantation? A potential mother cannot require that a tiny clot of cells demonstrate its worth by swimming the Hellespont or slaying a dragon, but it certainly seems within the purview of biology to insist that any protobeing who threatens to be so demanding show its worth by excavating a place for itself and by doing so effectively, efficiently, and expeditiously. Otherwise, the menstrual flood next time.

A prediction that may be derived from the competence test hypothesis is that when pregnancy is out of the question, menstruation will not happen. This might mean that menstruation may be tied to sexual intercourse, which it isn't. Young women begin menstruating and keep doing so on their own schedule, not as a response to their first sexual experience. Nor does menstruation cease during a cycle in which no sexual activity takes place. However, amenorrhea is often experienced by women who are anorexic or who are extreme athletes, whose body fat has dropped so low that they are no longer ovulating. It makes adaptive sense: if you are not even going to produce an embryo, then why lose blood and tissue if there's nothing (regardless of its competence) to wash away?

Not so fast, however. Amenorrhea may be consistent with the competence test hypothesis, but on closer examination it doesn't really offer positive evidence in its favor because the proximate mechanism for normal menstruation is closely tied to ovulation: if ovulation is inhibited (because of illness, stress, low body fat, or something else), then the hormones that lead the endometrium to thicken and develop are missing, and so there is nothing to menstruate away. The argument from amenorrhea is merely an object lesson in cause and effect and in the seduction of confusing facts that are *consistent* with a hypothesis with those that genuinely *support* it, never mind prove it.

In conclusion, there is much to be said about the competence test hypothesis, including that it has problems! It isn't clear, for example, why it should be necessary to wash away the whole uterine lining just to pull the plug on an incompetent embryo. Why not just construct an endometrium into which only a well-built trophoblast can excavate successfully? Or why not delay a wee bit before giving access to those crucial nutrients that were produced during the secretory phase of uterine buildup? And if, as this particular hypothesis suggests, human beings have incorporated competence testing into their reproductive biology because each child brought into the world is eventually going to be the recipient of a heavy dose of parental investment, then what about other animals that also tend to put all their eggs in a few baskets? Why don't elephants or manatees or blue whales menstruate? Perhaps they have other mechanisms—biochemically subtle and not yet identified—to evaluate their embryos. Admittedly, these creatures don't expect to give annual birthday presents or to have to drive their young to soccer practice, but they all produce singleton offspring at intervals of several years. So they, too, should have been selected to ensure that potential progeny are up to snuff before they get to enjoy the rich harvest of a nourishing uterus.

Synchrony and Its Discontents

Finally, another menstrual mystery. It is well known—or at least, widely believed—that when women live together, whether in a dormitory, sorority, rooming house, bordello, nunnery, or prison, their menstrual cycles tend to become synchronized. Unknown, however, is the reason for this supposed coordination.

One "explanation" holds that ancestral women's menstrual cycles were keyed to the phases of the moon, specifically (for some obscure reason) ovulating together when the moon was full and menstruating when it was dark. Now, of course, we live bathed in—or buffeted by—artificial light, so any possible influence by the moon is greatly inhibited. In addition, women don't typically live in close proximity, as they supposedly once did, so they are less likely to engage in this form of sisterhood.

The lunar hypothesis is notably weak and seems to be based largely on the fact that menstrual cycles are about the same length as lunar cycles, which in turn is almost certainly a coincidence, albeit one that lends itself to romantic extrapolation. (Is it a coincidence, for example, that the human gestation period is nine months and that nine is about one-third the length—in days—of the menstrual cycle and that, moreover, three cubed is nine? Yes. And a silly, inconsequential, wholly manufactured coincidence at that.) A study of menstrual cycles among the Dogon of Mali, one of the "natural fertility populations" so popular among anthropologists, asked whether their menstrual cycles were correlated with lunar phases. Given that the Dogon do not have electric lights and spend most of their time outdoors, they would seem ideal subjects for lunar synchronization. There was none.[25] Not even a coincidence.

A better proximate explanation for menstrual synchrony can be found in pheromones, those almost mythic chemicals that one wafts into the air, or possibly deposits on surfaces or into natural orifices, to which others are sensitive, even though both sender and receiver are unaware of what is happening. Similar sex-relevant effects have been demonstrated in a variety of circumstances, including studies that have revealed, for example, preference for complementary gene markers (the so-called major histocompatability complex, MHC) among potential sex partners.[26] It has also recently been reported that heterosexual men respond preferentially and the sex-sensitive regions in their brains "light up" on magnetic imaging tests when they sniff the urine of women compared to that of men; interestingly, women and homosexual men show a comparable response to *male* sweat.[27]

The now classic report of menstrual synchronization is by psychobiologist Martha K. McClintock, who in 1971 provided the first empirical evidence in support of the theory.[28] Sometimes called the McClintock Effect in her honor, it is superficially similar to the previously identified Whitten Effect, by which the estrous cycles of female mice are synchronized by the presence of a male.[29] It is clear, however, that whereas the Whitten Effect is caused proximally by pheromones present in the urine of male mice,[30] women don't need a harem-keeping male in order to synchronize their cycles. If menstrual synchrony occurs wherever women are housed together in the absence of men, then presumably it must be achieved by some sort of female-to-female signaling, which is to say probably via pheromones.

In her original work, McClintock found that college friends living in the same dorm who started the school year with independent, unsynchronized cycles began menstruating two days closer together after four to six months coresidence. McClintock's study of 135 fellow students at Wellesley College became her senior thesis at Wellesley (by the way, McClintock graduated from Wellesley in 1969, along with classmate Hillary Rodham Clinton). Shortly thereafter, when she was a graduate student at Harvard, her results were submitted to the prestigious journal *Nature* at the urging of famed entomologist E. O. Wilson, whose work with ants had made him especially attuned to the role of pheromones.

Martha McClintock and her colleague Kathleen Stern later reported in a much-noted follow-up study that when women's underarm secretions were wiped under the noses of other women, the amount of LH released by the recipients' pituitary was increased, and the recipients' cycle length was reduced.[31]

What about the "why" questions, though? What might be the adaptive, evolutionary payoff of menstrual synchrony? If there were serious competition among women to provide adequate food for their offspring, then it would be advantageous for a would-be mother to *avoid* giving birth when others are doing so; this advantage should select for *asynchrony*, if anything. Alternatively, some species engage in "predator overloading," whereby everyone breeds at the same time so that would-be predators are quickly satiated and unable to take more than a small proportion of the food available: if a leopard can eat no more than, say, fifty pounds in a week, and if prehistoric human babies weighed on average ten pounds each, that's 5 babies every seven days. Extended over a year, that's 250 babies, a heavy toll indeed. But if all those 250 babies had been born in the same week, only 5 would be taken, and the chances of any woman losing her offspring if she had coordinated her cycle with the other 249 expectant mothers would be only 5 out of 250, or a manageable 2 percent.

The problem is that human beings don't give birth in population pulses like this, and there is no reason to think they ever did.

Another possible evolutionary explanation for menstrual synchrony is that cycling together should make male infidelity less likely because if most nearby women are fertile at the same time, men would presumably be more inclined to stay home so as to inseminate their "own" mates while also defending them (actually, defending themselves) against any possible reproductive poaching by unmated men. So perhaps menstrual synchrony is a kind of pro–pair bond, anti-infidelity conspiracy on the part of women—albeit a benevolent and unconscious one, favored by evolution because it promotes monogamy and is accordingly beneficial to women insofar as it keeps men nearby. (It is thus similar to the keep him close hypothesis for explaining concealed ovulation, which we discuss in chapter 3.)

The monogamy hypothesis for menstrual synchrony might apply to people living in pairs, but would work quite differently if our ancestors were like chimpanzees or bonobos, inhabiting multifemale, multimale troops. In this case, menstrual synchrony would seem to enhance female-female competition because even the most ardent males cannot copulate with more than one female at a time, and, moreover, females have evolved strategies of keeping males' attention beyond what is strictly needed for insemination. As a result, one can imagine that in the competition among females, the winner would be the first one to come into estrous in that she would "win" the attention of the first male to choose a mate, and he would presumably be the most dominant *and* the most desirable. The outcome might therefore be menstrual synchrony, with everyone pushing simultaneously and successfully to the front of the line. Instead of deriving from unintended womanly cooperation in support of the pair bond, menstrual synchrony would then have resulted from its opposite: female-female competition.

Or maybe coordinated cycling isn't competitive at all, but precisely the opposite—namely, a means of encouraging camaraderie by orchestrating a positive feeling of group coherence. Many women report, for example, that when they discover that several of them are "on their periods" at the same time, there is an enhanced feeling of collegiality, as though all are bonded by a common affliction—or, more neutrally, a shared experience. Ancestral women whose cycles didn't "go along" were possibly less likely to "get along."

Or perhaps synchrony has nothing to do with either competition or bonding, but rather is an example of the "wisdom of crowds," whereby aggregated information tends to be more accurate than that possessed by an individual.[32] In this view, if others are ovulating and getting ready to become pregnant, perhaps it's a good time to have a baby. Maybe those others know something—albeit unconsciously—that you don't. Keeping up with the Joneses might thus be more than simply a way of maintaining social status, but also a way of taking advantage of widely held information.

Many animals experience what biologists call "social facilitation," with individuals more likely to do something if others are also doing it: individuals often eat more, for example, when surrounded by others who are also eating. Courtship and breeding—at least among colonial birds—speeds up when the local population increases. Sluggish breeders can sometimes be erotically inspired by the sight, sound, and smells of others who are ardently occupied. Some time ago social psychologist Robert Zajonc showed that human beings, too, experience social facilitation, performing tasks more quickly, more successfully, and more frequently in groups than when alone.[33] Yawning and laughing are contagious (for reasons that are unclear); maybe ovulation is similarly subject to social influence because, for some reason, togetherness often trumps individuality.

But wait! Before generating additional hypotheses, we must take time to acknowledge a more fundamental uncertainty: it isn't even clear that menstrual synchrony is real.

There is no question that menstruation and the various other traits we explore in the book—concealed ovulation, the presence of prominent nonlactating breasts, orgasm, and menopause—exist. What unites them, in addition to their provenance among women, is the mystery of their evolutionary "why." At present, however, what is truly enigmatic about menstrual synchrony is whether it occurs at all. Perhaps it is only an urban myth.

Part of the problem may be statistical. In her now classic study, Martha McClintock found a five-day difference in the menstrual cycling of cohabiting women, which, given an average twenty-eight day period, would seem very meaningful. But it isn't a question of five days (somewhat synchronized) versus twenty-eight (unsynchronized), but rather five days versus seven. Think of it this way: if two women with independent twenty-eight-day cycles began menstruating on the same day (zero difference between them), that would be full synchronization. However, their cycles can't be more asynchronous than to be fourteen days apart because even if they were separated by, say, fifteen days, they would actually be only thirteen days distant once they start anew. Thus, the average menstrual separation expected based on chance alone—and without any synchronization—would be seven days, the midpoint between zero and fourteen. This makes McClintock's finding of five days much less impressive than meets the eye. After all, half of the time, two independently cycling women should, by chance alone, be "synchronized" *closer* than seven days! And five is pretty close to seven. It's the job of statisticians to figure out whether such a difference is likely to be meaningful, which depends among other things on how large is the sample in question. In fact, statistical arguments have already been lined up on both sides.

Adding further complication is the fact that because menstruation not uncommonly lasts four or five days, there is substantial opportunity for any two women to experience overlap. As a result, the mere existence of overlap, as distinct from simultaneous onset, is not necessarily evidence for synchronization.

Yet more uncertainty is introduced by the difficulty of defining menstruation; for some women, it begins with an obvious flow and concludes with equally dramatic cessation. But menstruation isn't necessarily like turning on a faucet. It often starts and ends with inconspicuous spotting, making it difficult for people to agree on the exact days involved. Even if menstrual synchrony turns out to be genuine, it certainly isn't what statisticians call a "robust finding," but rather at most a matter of a few days more or less. Accordingly, the fact that those few days can be devilishly difficult to specify makes things yet more obscure. Putting numbers on so elusive a phenomenon may be a case of what philosopher and mathematician Alfred North Whitehead called the "fallacy of misplaced concreteness," or the process of regarding something abstract as a material entity simply because it gives the illusion of being measurable.[34]

Add the facts that a study of lesbian couples failed to show menstrual synchrony[35] and that the most reliable research on menstrual cycles in a nontechnological population, involving those lunar-insensitive Dogon, found no evidence for menstrual synchrony,[36] and the notion of menstrual synchrony itself threatens to tumble down like a house of cards.

For now, the safest generalization about menstrual synchrony is simply that the scientific jury is out. Considerable arguments exist on both sides, and— like the best of mysteries—a resolution is not yet in sight.[37]

So why do women menstruate (whether they do so in synchrony or not)? We don't know; it is an enigma. Period. But although *why* remains a secret, we know a great deal about *how,* the intimate connection between menstruation and ovulation. Every young woman whose entry into adulthood is marked by "getting her period" does so because she is beginning to ovulate and in the process is entering into one of the deepest and most hidden of all female mysteries. And this, in turn, leads to a new array of stories.

Here is John Berryman's account of an attractive woman sitting near him in a restaurant:

Filling her compact and delicious body
with chicken paprika, she glanced at me twice.
Fainting with interest, I hungered back
And only the fact of her husband & four
    other people
kept me from springing on her.[1]

Nothing much happens (no one springs on anyone) except that Mr. Berryman concludes his meditation by raising a show-stopper question, which is also the key inquiry of the present chapter: "What wonders is she sitting on, over there?"

Even if Berryman's ephemeral inamorata were totally naked, the poet would *still* have reason to wonder about what is probably the greatest female secret of all time. It has nothing to do with external genitals, which, clothed or not, aren't especially puzzling—or, at least, they aren't immersed in biological mystery. Rather, what is really important and altogether enigmatic about the woman who captures Berryman's attention (and, indeed, about any woman) is this: the status of her ovaries.

When people indulge their more carnal imaginings, they don't generally spend much time contemplating that which is deep inside: her fallopian tubes/his epididymis, her uterus/his prostate, or her eggs/his sperm. Another thing people don't normally think about but that warrants attention is the question, Why don't women develop bright pink bottoms when they ovulate?

Of course, whether thought about or not, most women ovulate just fine even when the evidence is

3
_____

Invisible
Ovulation
_____

kept chastely under wraps. Once a month or so, a ripe egg emerges from an ovary and eventually makes its way down the fallopian tubes to be fertilized. Or not. But whatever the ultimate outcome, it all takes place under the tightest secrecy. In fact, Mr. Berryman and his ilk probably don't even realize what wonders they don't know about.

Most female mammals are completely above board when their eggs are ripe and ready to encounter a willing sperm: their behavior changes, they may show physical signs of genital enlargement, and they emit characteristic chemical signals—pheromones—as well. This is why male mammals spend a great deal of their time sniffing the females' genitals, especially when the latter are in "season." Stallions, for example, also make a characteristic facial expression with a wrinkled nose and protruded lip when they detect estrous odors. Called *flehmen,* the expression looks as though the stallion has just smelled something awful (chances are, however, that he would describe the experience quite differently). Mares in heat oblige by presenting their rumps to be sniffed and then produce a vast amount of odoriferous urine that leaves no doubt in anyone's mind—even the human observer, comparatively obtuse in such matters—that they are ready to mate.

Among primates, too, the signs are often undeniable. Visit the monkey house at a zoo and you'll likely be convinced that monkeys or chimpanzees by and large provide not only chemical indicators as to their state of sexual receptivity, but also visual and even tactile cues. No one—not even a human being and least of all a sexually aroused male chimp—can miss the red, engorged rump of an ovulating, estrous female. Even more obvious than the proverbial nose on her face is that bright pink swelling on her rump.

Yet few people can tell when their friend, neighbor, relative, wife, or lover is ovulating or about to. Even more perplexing: If you are a woman, can you tell when (or if) you've done so yourself? Why is it such a covert operation for human females?

To be fair, this is a bit of an exaggeration. *Homo sapiens* is not the only primate species in which ovulation isn't publicized. Gorillas and orangutans, for example, don't offer up any external cues; nonetheless, their behavior changes when their eggs are ready to receive a spermatic visitor—that is, when the females are ready to conceive—leaving no doubt in their male consorts' minds. They also broadcast pheromones that are easily detected, at least by other gorillas and orangutans. It is one thing, however, to be stylishly subtle, quite another to be downright secretive, even deceptive.

It must also be pointed out that some women can in fact detect their own

Stallion showing characteristic "flehmen" face. Photograph by Risto Aaltonen.

ovulation by a pain—often sharp or cramping, but rarely severe—known as *Mittelschmerz,* "middle pain," when their egg is released (*middle* because it occurs midway through the menstrual cycle, not because it takes place in the middle of the body, although it does). It can arise on either the right or left side of the abdomen, depending on which ovary has just ovulated. As we'll see, subtle behavioral changes also surround ovulation, although most people don't know *Mittelschmerz* from Middle Earth. A study titled "Validating Signals of Ovulation: Do Women Who Think They Know, Really Know?" examined hormone levels in thirty-seven women who claimed that they could detect their own ovulation. Even in this self-selected group, the subjects were correct less than half the time, leading the authors to conclude that "for most women, ovulation is concealed."[2]

It is possible to pinpoint ovulation by careful daily temperature taking and by close attention to changes in the cervical mucous, as women seeking to get pregnant are instructed to do. But these methods are difficult and not completely reliable. In fact, it is a remarkable and telling fact that even now

Adult female chimpanzee with characteristic sexual swelling. Photograph by Ian Gilby.

in the technoadvanced, biomedically sophisticated twenty-first century, and despite intense interest and genuine consequence, there is still no simple, reliable "rhythm method" of birth control; that is, there is no guaranteed way for women to know with confidence when they are *not* fertile because, conversely, there is no easy way to know when they are. At the same time, pharmaceutical companies make huge sums of money marketing test kits that provide women with precisely the same information that most mammals get for free. (Women's ovulation is so secret that as recently as 1930, misinformation was rampant even among physicians and biologists as to when women were most likely to conceive: some felt that it was equally likely at any time in the menstrual cycle. Others—incredibly by modern standards—actually believed that fertilization was most likely around the time of menstruation.)

Before going any further, however, we must confront—and not for the only time—evolution's null hypothesis: that perhaps the trait in question hasn't really been selected for at all. Another way of putting it: maybe the adaptive, selected-for trait in this case isn't ovulatory concealment à la human beings, but conspicuousness à la chimpanzees. And here, once again, uncertainty abounds. Neither we nor anyone else can say for sure whether concealment or conspicuousness is the default biological setting. In short, maybe human ovulation isn't actively concealed, but simply inconspicuous, which is to say there is no guarantee that concealed ovulation in human beings is the result of positive selection pressure for this outcome, whereby our ovulation-concealing ancestors left more genetic representatives than did the ovulation displayers. Just as the human embryo develops female anatomical traits unless it possesses a Y chromosome (that is, morphological femaleness is the default condition), perhaps the understated pattern characteristic of human ovulation is the undifferentiated, unselected biological matrix, whereas the shout-out-loud, Technicolor ovulation of chimpanzees is the one that is "derived" and that primarily needs explaining.

We think this explanation unlikely. For one thing, nearly all mammals announce their ovulation, albeit not always as loudly as our closest relatives the chimpanzees and bonobos. By itself, this pattern is strong presumptive evidence that early hominids were relatively uninhibited about announcing their ovulatory status and that concealment—because it is so rare—is a more recently acquired trait; hence, it is likely to be the result of positive selection pressure that generated it.

In addition, there is essentially no variability when it comes to clear anatomical demarcation of human ovulation; in plain English, no women advertise their fertility in anything like the chimpanzee style. All women are ovulatory concealers, which, in turn, argues for a strong adaptive value in doing so. Consider, by contrast, a trait that is nonadaptive (or, at most, under very weak selection pressure): eye color. There is little reason to think that brown eyes render their possessors more fit or less fit than do blue eyes or green or hazel. Indeed, the human population is composed of people with brown, blue, green, and hazel eyes. If everyone had the same eye color, it would seem reasonable that this color was selected for or, at least, that departures from it were selected against. By the same token, if ovulation were concealed among *Homo sapiens* because of a kind of inertia or anatomical laziness in which selection hasn't bothered to promote anything discernible, then it seems likely that some women, somewhere, at some time in recorded history would have found themselves with conspicuous ovulation. Just as the diversity of human eye color suggests that eye color itself is *not* a clear-cut adaptation, the lack of diversity when it comes to concealed ovulation argues that such concealment probably *is*.[3]

It is also worth noting that women often receive substantial—and unappreciated—advance notice before getting their period, even though the regular onset of menstruation is heralded by fewer hormonal changes than is ovulation and, moreover, is much less biologically relevant to the rest of one's life. Put this all together, and it seems reasonable (although not guaranteed) that concealed ovulation isn't just a passive human trait, but one that has been produced by natural selection. In other words, the likelihood is that ovulation isn't just hard to detect; *it is actively hidden* by a woman's own body. But why?

Why should human beings partake of their own conspiracy? If nothing else, it would seem that any woman who knows when she is fertile and makes no bones about it would do a better job getting fertilized on schedule or avoiding pregnancy, whichever she desires. Wouldn't honesty be the best policy? Probably—unless there is good reason to lie.[4]

Keep Him Close and/or Guessing

One argument for ovarian dishonesty is that concealed ovulation keeps the male around. In other words, it plays an important role in pair bonding. According to this view, women hide their ovulation because that's how our first grandmothers kept our first grandfathers around in a kind of sexual thrall. The presumption is that sex is the glue that keeps men and women together; if women were sexually interested—and interesting—for only a few days each month, pairs wouldn't stick together the rest of the time because men would be off looking for additional girlfriends.

This statement assumes, among other things, that if ovulation weren't concealed, then sexual receptivity would be limited to the time of egg release—not a bad assumption because among those mammals in which ovulation is conspicuous, females' interest in sex runs exactly parallel to their estrous state. We don't mean that sexual swellings *can't* be decoupled from estrous behavior: it would seem biologically possible, at least, for women to be continuously receptive throughout their cycle and still advertise egg release during the limited time that it occurs.

One study examined sixty-eight different primate species, looking for correlations between visible signs of estrus and the mating system, whether monogamous, harem forming, or promiscuous. The results were confusing: not a single monogamous species turns out to be a conspicuous ovulator, which argues in favor of the keep him close hypothesis. At the same time, however, concealed ovulation does not guarantee monogamy; some sexually "cryptic" species, such as vervet and spider monkeys, aren't monogamous at all.[5] And to human observers, other species may appear to hide their ovulation, but in fact this impression is more a function of human olfactory obtuseness. Recent research on Barbary macaques (sometimes called Barbary "apes," although they aren't apes at all), has found that in this promiscuously mating species, males are especially likely to solicit sex from females who are ovulating, showing that insofar as male Barbary macaques are concerned, the timing of female ovulation—obscure as it is to the human observer—isn't concealed at all.[6]

Moreover, it seems that in baboons, at least, not only is ovulation announced publicly, but the degree of anogenital swelling accurately reflects a female's reproductive prospects, to which male baboons in turn are sensitive. In a paper titled "Sexual Swellings Advertise Female Quality in Wild Baboons," Leah Domb and Mark Pagel found that

[f]emales with larger swellings attained sexual maturity earlier, produced both more offspring and more surviving offspring per year than females with smaller swellings, and had a higher overall proportion of their offspring survive. Male baboons use the size of the sexual swelling to determine their mating effort, fighting more aggressively to consort with females with larger swellings, and spending more time grooming these females. Our results document an unusual case of a sexually selected ornament in females, and show how males, by mating selectively on the basis of the size of the sexual swelling, increase their probability of mating with females more likely to produce surviving offspring.[7]

Even those nonhuman female primates that *don't* show conspicuous anogenital swellings aren't necessarily concealing their ovulation; their reproductive state is typically quite apparent to males.

By contrast, the mystery isn't that human beings—unlike baboons—simply refrain from signaling their degree of fertility by how inconspicuously they advertise their ovulation. Rather, it's that human ovulation itself is Top Secret. So let's pursue for a moment the argument that concealed ovulation may have developed in the service of bonding early male-female *Homo sapiens,* uniting them not just in the physical act of love, but also as social partners. If, alternatively, prehistoric women's sexual receptivity had been limited to just a few days each month (presumably, at ovulation), male-female associations might then have been comparably limited and transitory, as it is among baboons and macaques—and, indeed, among most mammals, whose males contribute little to rearing their offspring. According to the keep him close hypothesis, prehistoric women, by concealing ovulation and at the same time expanding the duration when they are sexually receptive, may have enlisted sex in the service of male-female partnership. These adaptations would meet an important evolutionary goal: keeping a man nearby, available to assist the woman as well as her/their children.

Such a strategy would presumably have had enormous significance for *Homo sapiens* because, unlike other primates, human infants not only are altricial, or helpless after being born, but remain profoundly needy for a long time. This neediness in turn likely conveys a substantial benefit to having two parents, not just a genetic pair (everyone has that), but a devoted duo that helps rear them, defends them against predators, brings home the mastodon steaks, helps keep the cave bears at bay, drives the car pool to soccer games, and pays the orthodontist.

This explanation, although superficially plausible, may be less convincing than meets the eye. In particular, if we look at the animal world, it is clear that daily sex isn't necessary for a cozy one-to-one parental relationship or even an intense one. Many bird species are socially monogamous, with male and female closely associated and maybe even "in love," despite only copulating rarely and briefly when conception is most likely. Gibbons commonly go several years without sex, yet they remain reliably pair bonded (more or less). Most of us know people who stay devoted to each other even though the sexual spark may be only intermittent, if not altogether extinguished. And some marriages fail despite intense and satisfying sexual chemistry; indeed, sex is sometimes the only thing that works in a relationship, and if so, it is rarely enough to keep the pair together.

It seems likely, nonetheless, that regular and satisfying sex can help keep a couple together even if carnal satisfaction is neither necessary nor sufficient to guarantee perpetual bliss. And this likelihood, in turn, makes it intuitively persuasive that women, by hiding their ovulation, may succeed in making themselves sexier than they would otherwise be and that men respond by hanging around more than they otherwise would. But as useful as it may be to keep a man around, it may also be handy to keep him just a wee bit uncertain.

And so a variant on keep him close centers less on the woman and her receptivity than on the man and his interest in her. The idea is simple enough: what he doesn't know won't hurt him and may even help her. Richard D. Alexander and Katharine M. Noonan first broached this idea nearly thirty years ago, and it remains a cornerstone hypothesis: if men don't know when their mates are fertile, they may be biologically obliged to stay around all the time. We call it the keep him guessing hypothesis, a kind of sexual shell game. If any given day of sexual intercourse is neither more nor less likely to produce a child than is any other day, it might behoove a male to stay with "his" female in order to enhance his reproductive prospects.[8] His uncertainty (is she or isn't she?) becomes her certainty (she'll have him nearby whether she conceives or not). At the root of this notion, then, is the suggestion that a man might stay with a woman because, in effect, he wants (albeit unconsciously) to monopolize her eggs and can't do so except by playing a very inefficient shell game: turning over each opportunity, day after day, not knowing which has the prize and therefore obliged to keep trying because the woman keeps it—the elusive egg—under wraps.

This prospect—that women's ovulatory secrecy is a biological ploy to keep men hostage, playing to their uncertainty rather than to their love—admittedly doesn't make the heart sing, but a lack of sentimental appeal doesn't make this hypothesis or any other less likely to be correct. Whereas Rudyard Kipling's Just-So Stories were celebrated for their charm, evolution's stories must be judged on their plausibility and even more on their evidence.[9]

Unfortunately, however, for anyone seeking certainty at this time, the evidence is contradictory, at least regarding our nearest primate relatives. Among chimpanzees and bonobos, whose ovulation is exhibitionistically displayed, although individual males are extremely attentive to ovulating females, for the most part they either cannot or do not monopolize sexual access at such times. Instead, these species experience something of a sexual free-for-all, with swollen females mating about six to eight times per day; Jane Goodall recorded one female who mated fifty times in one day! There is typically little obvious competition among the males, nearly all of whom mate with a swollen female. But this doesn't mean that male chimpanzees necessarily share their sexual partners. In another behavioral pattern, a top-ranking male will sequester a female, typically at the height of her fertility, and form a consortship during which he does, in fact, pretty much monopolize her mating. The situation is therefore murky, rendering it at least possible that if human beings were overt ovulators, like chimps, they would be less inclined toward pair bonding. Or (as we'll see later) maybe the opposite is true: perhaps because humans are inclined toward pair bonding, they have evolved to suppress clear signs of when they are ovulating.

It is quite clear that at peak fertility chimpanzee females are especially choosy, with persistent preferences for the same males, presumably those who maximize the fitness of the female's offspring and thus of the chooser in question. At the same time, male chimps are especially likely to mate with females who are ovulating. In one study, researchers observed 1,137 copulations among chimpanzees, of which all but 32 involved a female whose sexual skin was maximally swollen.[10] If women were more "chimpy," with a limited number of obvious, ovulatory days, it is certainly possible that men would copulate with them at such times, then seek other women in similar condition, bolstered by the added confidence that their mates (now nonovulating and unreceptive) would not cuckold them in the meanwhile.

So let's grant that prehistoric men, whose women were good at keeping their ovulation secret, may have been especially inclined as a result of this secrecy to keep their women company. Being at least somewhat monogamous and largely monopolizing their mate's sex life, such men would have had a fairly high level of confidence that their children were in fact theirs. (Women don't have this problem, of course, because they cannot be deceived as to whose genes reside in their own children.) Ironically, therefore, women may have biologically and benevolently outwitted men by keeping men ignorant as to their ovulation and thereby inducing them to keep regular attendance, the result being that presumed fathers became *more* confident as to their paternity and therefore more likely to act as fathers, not just inseminators.

At the same time, the interests of male and female do not perfectly coincide, which sets up an interesting possibility: a coevolutionary arms race between women who were presumably selected to conceal their ovulation and men who were under strong selection pressure to penetrate the disguise and determine when ovulation was actually taking place. After all, in nearly every other mammal species, males seem to be well informed as to when a female is ovulating.

Although it is clear that concealment has been victorious at the level of conscious human awareness, there are tantalizing hints that at least to some extent men have not been left entirely clueless. Devendra Singh and P. Matthew Bronstad persuaded eighteen female college students to sleep alone in the same T-shirt for several days during their follicular/ovulatory phase and then in another shirt for several postovulatory days. When fifty-two young men were then asked to smell the eighteen pairs of shirts, the odor of the shirt worn during the follicular phase (just prior to ovulation) was preferred in 75 percent of the cases.[11] Moreover, another study shows that this preference disappeared when the women had been taking hormonal contraceptives, which suppress ovulation.[12]

These results strongly suggest that ovulation may not be perfectly concealed even in human beings. There remains no doubt, however, that it is superficially hidden, at least to our conscious minds. It is also possible that although ovulation is not publicly announced, once men are familiar with a partner and her odors, they can unconsciously tell whether she is ovulating or not even if they don't know what the different smells actually mean and even if they can't do the same with strangers. If confirmed, this notion would support the keep him close hypothesis, although it wouldn't necessarily refute other explanations of secret ovulation. An expanding list of studies seems to suggest, in any event, that although women, unknown to themselves, are adapted to obscure their ovulation, men are being selected to detect it.

It is also noteworthy that pair-bonded men act differently toward their sexual partners—engaging in behavior more suggestive of "mate guarding"—when these women are in the fertile phase of their monthly cycle,[13] although, once again, the men are not consciously aware of what they are doing or why. Not only that, but when both men and women were asked to judge photographs of women's faces taken during both the fertile and the infertile phases of their menstrual cycle, they considered the former more attractive than the latter.[14]

Thus far, we have two possibilities for the adaptive significance of concealed ovulation, both assuming that it evolved as a way to keep a man around: either by keeping him sexually satisfied or—in a sense—by taking advantage of his sexual (more accurately, his genetic) insecurity.

Ironically for the keep him close hypothesis—which claims that concealed ovulation might have been selected to encourage monogamy on the part of men—concealment could also have had the opposite effect with respect to those women who were successfully secretive: namely, facilitating sexual liaisons with a variety of different men. After all, it would be relatively easy for a woman's mate to defend her from other suitors if her time of sexual receptivity were limited and also clearly demarcated because that would be the only time such "mate guarding" would be strictly necessary.[15] It is more difficult, by contrast, to guard her all the time, which is called for insofar as women don't reveal when such guarding is especially needed. Concealed ovulation is intimately connected to another unusual characteristic of women: their lack of estrus. Among most mammals, estrus or "sexual heat" is closely attuned to egg release; thus, to say that the latter is hidden is to say that there is no time when female sexual receptivity is especially intense or in any way out of control.

It is difficult to get good evidence as to whether women have more sex or less when they are fertile and especially hard to know whether any difference between fertile and nonfertile frequency is due to biology or to conscious choice. After all, women who want to get pregnant should intentionally have more sex when they are fertile; in contrast, women who want to avoid pregnancy should have less.

In any event, there is no question that on balance more human copulations take place when a women is not fertile, if only because there are many more such days, and to a large extent human beings engage in sexual intercourse throughout the cycle.[16]

This fact leads to several more possibilities. Mark Twain once announced that it was easy to stop smoking: he had done it hundreds of times! Similarly, as we shall see, it is easy to offer an explanation for concealed ovulation; many people have done so in many different ways. What's difficult is deciding which explanation, if any, is correct. For a definitive answer, it would help if we had two different subsets of women: some whose ovulation is overt, and others whose ovulation is concealed. Then we would be able to compare their own and their men's behavior and possibly reach some firm conclusions. As it is, though, we can only use logic, apply it to the available evidence, and leaven any frustration with the intrigue and satisfaction that attends a good mystery story, especially one that has numerous Just-So components.

Infanticide Insurance

First, a grisly fact: infanticide is common among animals, and most of the time the perpetrators are adult males. Even more deplorable, when it happens, there is generally a method to the males' mayhem: an ascendant male replaces the dominant harem-keeper, whereupon that new alpha proceeds to kill the existing youngsters. (This doesn't happen all the time, but often enough that it is clearly something real and not just an occasional aberration.) It is a pattern that occurs among lions, rodents, and even certain primates, including the gray langur monkeys of Asia, among whom the phenomenon was first reported in pioneering research by primatologist-anthropologist Sarah Hrdy.[17]

The logic is straightforward, gene-based selfishness. When a new male supercedes the old harem-keeper, he finds himself surrounded by adult females with whom he is ready and eager to reproduce. In fact, that is why the newcomer worked so hard to supplant the previous dominant male: selection favors males who succeed in projecting their genes into the future. More accurately, it favors those *genes* that succeed in getting themselves projected into the future through the actions of the bodies in which they reside; in this case, the result is the same.

Sometimes, however, the newly ascendant male is thwarted, at least temporarily, if one or more of the females are lactating mothers because nursing females in general don't ovulate. In such cases, it can be advantageous for the male to dispose of any nursing infants, after which the mothers typically resume cycling and proceed to mate with their infant's murderer. (Evolution definitely does not offer an attractive model for human behavior.)

Recall that in such cases the young victims were sired by the preceding harem master. This point is important because it means that infanticidal males are under few inhibitions considering that the genes they are destroying came from someone else.

Add another important fact—a father's tendency to tolerate his own children, which is also a result of selection—and we can glimpse how this system generates yet another possible explanation for concealed ovulation. Hrdy, who had awakened the scientific community to the reality of infanticide, was the first to propose this idea, too. The strategy is simple: females should have sex with a variety of different males. By so doing, they aren't being promiscuous—that is, indiscriminate as to their sexual partners. Rather, they may be highly discriminating, going out of their way to copulate with *different* males so as essentially to take out an anti-infanticide insurance policy for their offspring. If there is a changing of the guard—and eventually, there always is—the new harem master may spare his old flame's offspring out of genetic selfishness rather than sentiment.

In a study of free-living langur monkeys, researchers found that females have extended timeframes during which they are sexually receptive even when not ovulating. They also discovered that even though dominant males tend to monopolize receptive females, nondominant males father a substantial proportion of offspring, which, in the authors' words, provides the "first direct evidence that extended periods of sexual activity in catarrhine primates [Old World monkeys] may have evolved as a female strategy to confuse paternity."[18]

This explanation assumes that if a female had an obvious period of gaudy, pink-rumped estrus, a small number of males would likely attend her closely, making it difficult if not impossible for her to spread her sexual favors and thus to generate an array of deceived "fathers." As such, the infanticide insurance hypothesis is almost the exact opposite of keep him guessing because it suggests that females obscure their ovulation in order to reap the benefits of being more sexually free and copulating with a number of different partners, whereas keep him guessing argues that keeping one's eggs under wraps is a ploy to keep a specific male in regular attendance, which would seem to make it *less* likely that the female can have a fling with someone else.

Things are yet more complicated, although not impossibly so. Infanticide insurance is only the most dramatic potential payoff that prehistoric females might have derived from mating with more than one male. We are pleased to acknowledge that males can do more than refrain from committing infanticide! They can share food, provide protection from predators, assist in rearing offspring, and so forth. It has been established that among many animal species, females gain material resources and other such benefits from "extrapair copulations" (those with individuals outside the social pair bond). There is no reason why similar payoffs could not have accrued to ancestral humans, too. In fact, there is abundant evidence that monogamy is a derived condition among human beings, something of which our species is capable and that in many circumstances is even highly desirable, but that is nonetheless "unnatural."[19]

Keeping Control

Among the payoffs to nonmonogamy is an especially subtle one—namely, obtaining good male genes to combine with one's own. Unlike most other primates, human beings profit mightily from paternal care, which, in turn, seems to be one of the major considerations responsible for human monogamy and perhaps its greatest payoff. But social monogamy must be distinguished from sexual monogamy because the evidence is overwhelming that *Homo sapiens,* like most other species, is no stranger to adultery.[20] At the same time, the benefits of paternal assistance (to offspring and thus to the mother as well) are so great that ancestral female hominids would probably have been ill advised to behave like chimpanzees and engage in public sex with numerous males, which almost certainly would have made it very unlikely that their identified social mate would invest heavily in the offspring that resulted. The prospect therefore arises that via concealed ovulation, women discovered how to have their cake and eat it too: obtain desirable genes (and possibly other resources as well) from males with whom they mated on the sly, while also retaining parenting assistance from their socially designated partner. By obscuring their time of maximum fertility, women would therefore have gained control over their reproduction insofar as they succeeded in duping their "mate," thereby obtaining both a dish-washing, diaper-changing, bill-paying husband *and* a sexy lover whose good genes would grace her offspring.

Moreover, by abandoning estrus, women may have literally gained greater control over their choice of sexual partners. Consider the fact that in many species, females in heat are no more rational than their male counterparts. Without estrus, women are able to stay comparatively cool and in sexual control, albeit nonetheless passionate when appropriate. According to this particular argument, then, by foregoing estrus, women have become masters of their genetic fate, empowered to pick and choose, deciding (maybe not consciously, but with a degree of judgment nonetheless) among potential suitors. The word *estrus* comes from a Greek term for a parasitic fly that pursues cattle and drives them crazy, and, after all, a female animal in estrus does seem more than a little crazy. By the same token, a female who is not in estrus—which is to say, all women all the time—is more likely to be sane, sober, and better capable of good judgment.

In his poem "If," Kipling wrote about the merits of being able to keep your head while all those around you are losing theirs. If you can do this, according to his famous poem, "then you will be a man, my son." Maybe by favoring those women able to "just say no" to the tyranny of estrus, natural selection endowed our female ancestors with the ability to reap the benefits of being more discriminating than any other animal (and probably more so than most men, to boot). "If you can keep your secret," evolution might have been saying, "while all those around you are divulging theirs, then you will be a reproductively successful woman, my daughter."

It may be significant that among nonhuman primates, ovulation is prominently announced among those species—baboons, macaques, chimpanzees, bonobos—that live in multimale groups in which many different males typically mate with an estrous female.[21] Interestingly, in such cases, males tend to have unusually large testicles, which produce vast amounts of sperm because the actual competition among males takes place largely within each female's reproductive tract. Females among these species do not appear to exercise much sexual choice, although some primatologists—including, notably, a number of prominent women scientists—have argued that in fact the females of these species are indeed demonstrating a choice: to mate with many different males!

It is also noteworthy that while mating, females of these species often vocalize loudly, which attracts the most dominant males, and, of course, the very fact that these females are being so gaudy strongly suggests that they are competing anatomically and physiologically for the attention of the best sperm donors. By contrast, those of our immediate human ancestors who by virtue of their genetic makeup followed a different tactic and turned down the sexual heat by concealing their ovulation may well have been favored by natural

Evolutionary Hypotheses Explaining Women's Concealed Ovulation

Inconspicuous: not really concealed
Keep him close: provide sexual satisfaction
Keep him guessing: sexual shell game
Infanticide insurance: confuse paternity
Keeping control 1: choose sire
Keeping control 2: obtain resources
Keeping the peace: minimize chaos
Competition avoidance: success of the secretive
Lioness strategy: exhaust the male
Covering tracks: facilitate infidelity
Headache: consciously avoid pregnancy

selection if by obscuring their reproductive status, they kept their mates less fervent and thus more tractable.

Think of the situation for women this way: if women were sexually receptive for only a day or so each month, at which time they broadcast their availability by sudden, seductive signals of sight, smell, and sexual interest, men might respond by huffing and puffing and blowing each other away—even more than they do now—leaving women little choice, perhaps, but to accept the victor. Dominant males may be desirable sperm donors, if only because their sons may also turn out to be dominant males (and their daughters more vital and successful), but they may also be terrible fathers, more interested in beating other males over the head than in caring for their children. By keeping their ovulation secret and thus dampening men's competitive ardor, women might have given themselves the opportunity of choosing partners who may be less pushy but more paternal.

At the same time, women pay a potentially devastating cost as a result of their concealed ovulation: it makes them more vulnerable to being raped. Among animals such as horses and dogs, males by and large are sexually interested only when females are receptive. Hence, mares and bitches are only rarely sexually attacked by stallions or dogs because when females aren't in estrus, males aren't interested, and when males are sexually demanding, females are too. A possible downside of women's having evolved to obscure their time of peak ovulatory receptivity may therefore be that they have ended up being more vulnerable to men's forcing unwanted sex upon them simply because as a result of concealed ovulation, men—clueless enough when it comes to romance generally—are even more clueless as to when women are likely to be biologically receptive. If so, this is hardly a matter of "keeping control," insofar as a case can possibly be made that as females evolved to gain control, some men (presumably those unlikely to achieve matings via courtship and female choice) evolved a tendency to engage in rape as a countertactic.[22]

Whereas sexual harassment remains a serious problem in the woman-made world of concealed ovulation, it does not require a highly developed imagination to picture how much worse it would be if brief periods of fertility were prominently advertised. It is therefore worth considering that sexually choosy women, whose hidden ovulations make them all the more capable of being choosy, are the ones who have inherited the biological world.

The keeping control hypothesis has two more variants. First, concealed ovulators probably cannot be guarded all the time, even by the most possessive males. As a result, ancestral women who did not broadcast their times of peak fertility might have been liberated to mate—albeit probably in secret—with the male or males of their choosing, not necessarily to garner infanticide insurance or additional resources, but rather to gain fitness-enhancing genes possessed perhaps by someone outside their mateship or even outside the immediate social group.

A growing body of evidence supports this idea by showing that women behave somewhat differently around the time when they ovulate. In a now classic Ph.D. dissertation examining sexual behavior among !Kung san ("bushwomen") of the Kalahari, anthropologist Carol Worthman found that libido is typically increased when fertilization is most likely and that at this time women are more likely to have sex with their husbands, but also with other men.[23] Similar findings have been reported for Western women. Along with enhanced sexual yearnings at midcycle, women are also more likely to feel self-confident and to report more orgasms during sexual activity, whether heterosexual or homosexual, involving either intercourse or masturbation.[24]

In fact, a number of recent findings point to subtle, previously unacknowledged differences in women's behavior at various points during their normal hormonal cycles. Thus, it turns out that when women are ovulating, they are especially likely to prefer images of men who look and sound especially "manly," to wear clothing that is sexier and reveals more skin, to speak with greater verbal creativity and fluency, to be more discerning when it comes to others' behavior, to be attended (guarded?) more closely by their male companions, to be perceived by both men and women as more facially attractive, to have a more acute sense of smell, and literally to be more restless and likely to move around more. The titles of some of these research articles provide a good idea of what they are reporting: "Women's Fertility Across the Cycle Increases the Short-Term Attractiveness of Creative Intelligence Compared to Wealth," "Ovulation and Human Female Ornamentation: Near Ovulation, Women Dress to Impress," "Preferences for Masculinity in Male Bodies Change Across the Menstrual Cycle," "Preferences for Symmetry in Faces Change Across the Menstrual Cycle," and "Menstrual Cycle, Trait Estrogen Level, and Masculinity Preferences in the Human Voice."[25]

Perhaps the most eye-opening of these research efforts has involved lap dancers, who—we are told—move around quite a bit, but to entertain men for money. Evolutionary psychologists Geoffrey Miller, Joshua Tyber, and Brent Jordan signed up lap dancers from nearby clubs and had them report on their monthly cycling: whether they were in high-fertility phase, luteal phase, or menstrual phase and how much money they were paid in tips at these different times. Eighteen women recorded their menstrual cycles and tips received during a sixty-day period. These women normally earned an average of $335 for a five-hour shift during their fertile phase, $260 per hour during their luteal phase, and a paltry $185 per shift during menstruation. As a kind of clincher, women using contraceptive pills showed no earnings peak, just as they had no fertility peaks. It isn't known whether ovulating women sent out pheromones or danced differently—presumably more seductively—but in either case men detected and responded to something about them.[26]

These and other results have led some evolutionary psychologists to propose that much of the conventional wisdom among reproductive biologists must be revised and that women do indeed experience genuine estrus, like other mammals.[27] Psychologist Steven Gangestad and biologist Randy Thornhill, for example, claim that women undergo a "dual sexuality" in which estrus is associated with ovulation and "extended sexuality" covers women's sexual behavior at other times, when conception is impossible. They argue that at peak fertility, women show a suite of behaviors that indicate a preference for male traits that suggest direct genetic payoffs ("sexiness") rather than those traits associated with long-term mateship (resources, cooperation, and so forth). According to Thornhill and Gangestad, "sire choice" operates at peak fertility, whereas "extended sexuality" functions to obtain nongenetic benefits from males: infanticide insurance, bestowal of resources, and so on. There is, indeed, mounting evidence for this distinction. As these two prominent scientists put it, "Whereas fertile-phase women were particularly sexually attracted to men perceived as arrogant, intrasexually confrontative, muscular and physically attractive, no cycle shifts were observed in women's attraction to men seen to be successful financially, intelligent or kind and warm. Men who appeared to be sexually faithful were less sexually attractive to fertile phase women; put otherwise, fertile women are particularly attracted to men who appear that they would not be faithful (probably because they possess features women find attractive in sex partners)."[28]

In addition, women who are paired with sexually unattractive men (as judged by these same women) find themselves more attracted to extrapair men, especially during their own fertile phase.[29] Not only that, but women who share comparatively many genes with their partners—and whose offspring are therefore liable to be somewhat fitness compromised as a result of this excessive genetic similarity—are more likely to become interested in extrapair men, whereas women who do not share many genes with their mates do not show a comparable extracurricular attraction.[30]

However, the question of "dual sexuality" is far from resolved. A massive study of the frequency of sexual intercourse, involving more than twenty thousand women in thirteen different countries, found no consistent changes associated with their cycles—except for a drop in sexuality during menstruation.[31] These women were no more likely to have sex when fertile (during their ovulatory phase) than when infertile (during their luteal phase).

These results argue for a single or unified sexuality rather than a dual sexuality in which women essentially seek fitness-enhancing sperm when fertile and a relationship-enhancing partner the rest of the time. Alternatively, it can be argued that women engage in sexual intercourse even when not fertile in order to obtain precisely those relationship-enhancing benefits and that their sexuality is thus still "dual," just motivated by different ultimate factors at different times.

For our purposes, the precise terminology is irrelevant, and, indeed, no one disputes the fact that women's sexuality is extended beyond ovulation and that female libido is to a large extent (although not completely) liberated from fertility. We agree, as well, that something hitherto unappreciated goes on among women such that during ovulation they are not only predisposed to act (typically unconsciously) in a way likely to get their eggs fertilized, but also better able—specifically when they are most fertile—to encounter and evaluate potential sexual partners.

These various midcycle changes are subtle, however; otherwise, everybody would have known all along that women experience estrus, so the fascinating flurry of findings cited here wouldn't have surprised anyone. We conclude, therefore, that women do indeed experience concealed ovulation rather than true estrus because all evidence confirms that the peak of women's fertility remains hidden, at least in the sense that neither women nor their partners— whether through marital or extrapair interaction—are readily aware of what is going on. Human ovulation reveals itself as an inconspicuous "conspiratorial whisper" rather than as the obvious shout-out that occurs in chimpanzees.[32] Perhaps scientists should start talking about an oxymoronic "cryptic estrus" or use a similar neologism.

In any event, the ongoing discoveries that women show a sexually consistent suite of behaviors when they are most fertile comports with the keeping control hypothesis in that by concealing their ovulation—even as fertility is influencing their behavior—they ironically succeed in keeping control without being aware of the control they are exercising!

"In serially monogamous species such as ours," suggest the authors of the lap-dancing study, "women's estrous signals may have evolved an extra degree of plausible deniability and tactical flexibility to maximize women's ability to attract high-quality extra-pair partners just before ovulation, while minimizing the primary partner's mate guarding and sexual jealousy. For these reasons, we suspect that human estrous cues are likely to be very flexible and stealthy— subtle behavioral signals that fly below the radar of conscious intention or perception, adaptively hugging the cost-benefit contours of opportunistic infidelity."[33] In other words, fertile women can use their fertility to their own advantage.

The other payoff possibly connected with the keeping control hypothesis is that a nonestrous woman, instead of (or in addition to) keeping control of whom she mates with, may be more able to keep control of how she is repaid for granting sexual access to herself. Lap dancing isn't prostitution, but it shares some characteristics insofar as women trade sexual access (or at least stimulation) for money. As the authors of the New Mexico study note, lap dancers are not seeking mates, but money, and, moreover, "It seems that the optimal strategy for obtaining tips is to focus on men who are profligate, drunk, and gullible rather than [on] those who are intelligent, handsome, and discerning."[34] Men pay, and women get paid, even when the latter are not fertile. And it makes sense that women would be astute enough to employ their biological gifts toward that end, thereby besting men in a kind of evolutionary arms race. Even if no intercourse takes place, encounters of this sort are drenched in sexuality, powered by circumstances and cues that would certainly have been relevant to reproduction during most of our long evolutionary history.

Something closely akin to prostitution may also occur among other animals, with females exchanging sex (which they provide) for resources (which the males provide). Imagine a prehistoric female who sought to make a basic barter of sex (actual sex, not lap dancing) for food. Imagine, further, that her ovulations were conspicuously advertised, but that she wasn't ovulating at the time when a male had an especially appealing bit of fruit. In this case, the male might well refuse, depriving the female of considerable leverage; indeed, in most (but not all) mammal species, males will not copulate with nonestrous, nonovulating females. Not only that, but because the male would be most inclined to say yes when his chances of fatherhood were greatest, the female might well have been stuck with his genes when all she really wanted was his food.

In the alternative scenario, though, with her ovulation hidden, the same female could get the banana and eat it, too, while keeping her eggs untouched— all the while, moreover, giving the duped male the impression that perhaps he was also becoming a father.

To summarize and perhaps oversimplify a complicated hypothesis: maybe concealed ovulation evolved when women found it reproductively advantageous to increase men's confidence of paternity (and thereby increase the chance of their being good fathers), while at the same time retaining control over who really was the father of their children.

Keeping the Peace

Whatever else is true of human sex, and no matter how much it is publicly advertised, even sometimes flaunted, it tends to be *done* in private. And whatever else is true of concealed ovulation, it, too, is by definition private. So perhaps concealed ovulation is all about privacy—specifically, *keeping the peace.*

In short, evolution may have concealed female ovulation as a way of avoiding the chaos that might otherwise come with a public sexual free-for-all. Anthropologists Jane Lancaster and C. S. Lancaster asked what the world would be like if a woman's appearance and behavior accurately reflected her ovulatory status.[35] Consider the effect on, say, a typical male-dominated law firm were a female employee to show up for work with an unmistakable swelling. Even hidden under clothing, it would doubtless send out olfactory cues as well. Recall the poem by John Berryman that opened this chapter, in which—despite the presence of a woman's husband and four others, and with her ovulation chastely concealed—the poet claims he was barely able to restrain his lust. Now imagine that the woman in question is prominently and unambiguously divulging her ovulatory status. Every woman would need an armed guard (who in turn would presumably have to be a eunuch because otherwise, as Plato inquired in a somewhat different context, who would guard against the guards?).

Or picture a neighborhood high school packed with young estrous females and half-crazed adolescent males in the throes of near-terminal testosterone poisoning. Civilization, the argument goes, would be radically different, certainly more chaotic and maybe even impossible, if sexual receptivity were as obvious and compelling in humans as it is in chimps.

On a strictly biological level, however, the question becomes whether such disruption would be disadvantageous for the individuals involved. And here the keeping the peace hypothesis loses some of its luster. It is certainly possible that females have an interest in maintaining a modicum of decorum around the cave, the campfire, or the committee room, but this problem pertains mostly for the males. In short, selection shouldn't cause females to lose or acquire a trait just because it might benefit someone else. Evolution is about genes looking after themselves. In short, there must be a payoff for *the females in question* to keep such an important secret. Moreover, this payoff must benefit the secretive female—or, more specifically, her genes. A case can be made that when it comes to announcing one's reproductive availability, the announcer might actually benefit by the ensuing chaos because the fittest males would presumably rise above their competitors and mate with her.

If announcing their reproductive status were advantageous to women who are doing the broadcasting, such messages ought to be sent loud and clear for all to see and hear. Or smell or touch. But they aren't, presumably because it isn't—unless, perhaps, if done with subtlety, tact, and delicate evolutionary finesse.

Competition Avoidance?

Thus far we have considered efforts to explain concealed ovulation by focusing either on the loss of estrus (that is, the spreading of a woman's sexual receptivity throughout her cycle as opposed to the focused "heat" seen in some mammals) or on the hiding of sexual cues from men, or both. Here is another possibility: maybe concealed ovulation is a matter of women's hiding their precise reproductive status *from other women.* (To our knowledge, no one has yet proposed this.)

When biologists think competition, they usually think male-male, and for good reason. Throughout the animal world, especially among mammals, it is generally the males who are bigger, more lethally armed, and altogether nastier. Males typically are the ones with big racks of prongy antlers, massive horns, extralong canine teeth, and an inclination to make use of what they have. But this doesn't mean that females don't compete, too. In fact, students of animal behavior and ecology have recently begun to realize that females are also competitive, just more subtle about it.

And so it is that for years researchers believed that when they saw female monkeys gather round a newborn infant and pass the baby around, they were witnessing a kind of baby-sitting. Indeed, these additional baby-handling "caretakers" were labeled "aunties" until it was noticed that the purported baby-sitters were often doing just that: sitting on the babies!

In other species as well, females seek to improve their reproductive situation by worsening that of others. If they can't achieve the latter goal by being nasty to each other's kids, they sometimes interfere with mating opportunities. After all, what better way to diminish the success of a rival's offspring than by keeping her from having any? This, once again, is where concealed ovulation might come in, at least in the case of human beings. Let's assume that food is scarce, so that it is advantageous to female A if female B doesn't reproduce because it would mean more food available for A's own children. One possible option for A would therefore be to beat up B whenever B shows any sign of coming into season. (It has already been found that female-female aggression can inhibit ovulation.)[36] Or alternatively, A might keep B isolated, ostracized, or in hiding, far from the attention of any salacious males, so that even if B ovulates, neither she nor her would-be paramour(s) can do anything about it. Either way, under circumstances of this sort, it might pay female B to keep her ovulatory status secret.[37] By contrast, females who advertise their condition would be selected against, especially if they were socially subordinate to their reproductive rivals. The result would be concealed ovulation: the success of the secretive.

A plausible enough theory, but to generate confidence it would help to show that those species in which ovulation is not concealed are less troubled by intense female-female competition, and that concealed ovulation, for its part, is generally found among species in which female-female competition is otherwise intense. Thus far, sadly, no such correlation has been found.

Moreover, it would be encouraging if comparatively dominant and successful women, who have less to fear when it comes to female-female competition, turn out to be somewhat less prone to concealing their ovulation. A related prediction: if concealed ovulation is a strategy to diminish female-female reproductive competition, then women who are married or otherwise securely pair bonded might be expected to make less "effort" (physiological or behavioral) to conceal their ovulation. Once again, however, no research speaks to this question, largely because, as we have seen, ovulation is cryptic in all women.

The Lioness Within?

Female sexual desire is a kind of glass, either half-full or half-empty. The half-empty perspective points out that women don't experience estrus and, accordingly, their sexual drive is weak compared with their animal relatives (see the keeping control hypothesis). In the past, women's supposed sexlessness was trumpeted as only good and proper; sexual desire was something good girls didn't experience, and if they did, it wasn't proper to talk about it. In any event, they certainly didn't act on it.

But we can also look at women's sexuality as a glass half-full. After all, by the same token that women—unlike so many other animals—do not experience regular times of sexual frenzy, we can point out that they are in a sense sexually ready at any season, any time of their cycle, any time of the day or night. They are not sexually voracious during brief, identified times of "heat," but maybe that is just another way of saying that they are in heat, of a sort, *all* the time! Or at least always warm.

This view, in turn, leads to yet another possible explanation for concealed ovulation: call it the Lioness Strategy. Little known to most people, lions are among the world's sexiest creatures; they copulate upwards of a hundred times per day for the three days or so that the lioness is sexually motivated.

Some suggest that the female's insatiable demands make it unlikely her depleted mate will inseminate other lionesses, which in turn helps reduce the chance that her cubs will compete with other litters for limited food resources. Human females' remarkable sexual capacities—another way of construing concealed ovulation—may have evolved to serve an equivalent function: keep

Mating African lions. Photograph by Thomas Mueller.

a man satisfied or, better yet, exhausted, so he won't be inclined (or able) to inseminate anyone else.

Human beings are a sexually active species, although admittedly less so than the nearly insatiable lions. Maybe our great-great-great-grandmothers several thousand times removed found it in their interest to get in touch with the lioness within whenever our great-great-great-grandfathers developed an especially roving eye. This change by itself wouldn't necessarily produce concealed ovulation, but it may have given rise to women's ability and inclination to have sex whether ovulating or not—and to hide the fact that, at least in strictly reproductive terms, the sex might not be "genuine."

Expressing Love or Covering Tracks?

In our search for the underlying causes of concealed ovulation, let's take a minute to look at those animals that copulate a great deal because, as we have seen, one part of concealed ovulation is losing estrus and thus gaining the ability and inclination to have sexual intercourse throughout the cycle. Paradoxical as it may seem, loss of estrus can result in more sex, not less.

It used to be thought that human beings were unusual and perhaps unique in engaging in so much nonreproductive intercourse. After all, it isn't strictly necessary to copulate several times a week, on average, just to produce a child every few years or so. The argument went that among our own species, sex has been liberated from reproduction, part of the evidence being that in the rest of the animal world, by contrast, sex and breeding are tightly connected. We human beings had supposedly uncoupled our coupling from mere reproduction and began using it for "higher" purposes, notably love and bonding.[38]

This separation of sex and reproduction, in turn, might be part of the biological reason for concealed ovulation if it enables couples to bond together especially tightly, to express and enhance their love via "liberated sex." But, we might ask, *why* are sex and love so tightly connected? Why don't romantic partners express and enhance their love by picking each other's lice (many primates do just that) or via beautifully coordinated, mutually satisfying bouts of highly choreographed belching? One guess is that people find shared sex reinforcing to the pair bond for a particular reason. Hints of this reason can be glimpsed in other species.

The question is, Why do people and certain other animals copulate so often? The answer may seem distressingly practical, even cynical. When we look at those other species that have frequent sex, some even more than human beings, we find that many of them (bonobos, chimpanzees, some species of dolphins, and lions excepted) are "socially monogamous," but prone to being sexually *un*faithful. Certain birds—for example, northern goshawks, osprey, white ibises, and acorn woodpeckers—copulate hundreds of times for every clutch of eggs. And they don't limit those copulations to their designated partners.

The possible connection is as follows. When males are separated from their mates for a substantial part of each day, they are at risk of being cuckolded. In such cases, a sexually adventurous female would be risking little. After all, she is guaranteed to be the mother of her offspring, no matter how many partners she may have. The male, by contrast, has no such automatic confidence: "Mommy's babies, Daddy's maybes." Accordingly, males may be especially inclined to copulate frequently when they are at home to increase the chances that their offspring are in fact theirs. In turn, we might conclude that they "love" their mates all the more when given the opportunity to make love with them. Moreover, the more loving they do, the more love they feel, with "love" in such cases defined as a powerful inclination to remain with and be devoted to one's partner—because, in biological terms, the greater the confidence of shared genetic investment, the greater the love.

Now, look at the situation from the female's perspective. Let's say that the female osprey, northern goshawk, or white ibis is occasionally inclined to have sex with males in addition to her mate, perhaps because her nonpair partner is especially able or inclined to invest in her offspring or because his genes are particularly fitness enhancing. At the same time, however, she dearly wants to retain her social mate's parental assistance. It would make sense in such cases for the female to indulge her partner's sexual inclinations and to copulate often, if only because by assuaging his unconscious anxieties, she is more likely to obtain his continuing assistance and commitment while still remaining free to indulge her own extrapair inclinations when her mate is otherwise occupied.

Maybe all this has nothing to do with human beings, but it is probably no coincidence that other species among which mated pairs mate frequently are particularly likely to do so outside the pair bond as well. The prospect looms, therefore, that the remarkably high frequency of in-pair human sex isn't so remarkable after all, considering that human beings are more than a little prone to sexual dalliance.

So perhaps we don't advertise our ovulation because, at least in part, to do so would be to invite an unacceptably high level of sexual jealousy and obsessive mate guarding on the part of the male at these times, all the more so given that our own species is somewhat inclined to infidelity, which in turn suggests that women would be more fit if they didn't incite their social mates to keep close tabs on their sexual activities.[39] Lacking a single dominant peak of sexual desire and desirability, women are liberated to have sex with other males, not just with their designated mates. This possibility is a variant on keeping control (by keeping him guessing) insofar as it emphasizes women's sexual liberation, but it differs in paying attention to each woman's payoff, which needn't involve purchasing infanticide insurance or even sire choice, but rather covering her tracks. Not only that, but this liberation might also free a woman to have abundant sex with her designated partner as well, in the process keeping him sexually satisfied and less worried about his paternity. Perhaps less worried than he ought to be.[40]

The Paradox of Consciousness

Time now for a change of pace. All the hypotheses discussed so far try to explain why ovulation is concealed from others—especially from men, but also perhaps from other women. But why is this information kept from each ovulating woman herself? Why should women be in the dark about something so important going on inside their own bodies? As the king of Siam famously observed in the musical *The King and I*, "It's a puzzlement."

It isn't that people can't keep their own counsel and are somehow bound to be blabbermouths about whatever is happening within themselves. For example, everyone gets hungry, but this feeling is largely personal, an appropriately "inside thing." A person doesn't become bright purple or start twitching uncontrollably or give off a distinctive odor when he gets hungry. Similarly, if you develop an ulcer, it is *your* ulcer, not public knowledge, unless you choose to share the information.

Let's grant that there is likely some biological payoff to keeping one's ovulation secret from others, but why isn't every woman in on her own secret? Why the *self*-deception?

One possibility is that self-deception is necessary—or at least helpful—if one is going to deceive others. After all, the best liars are those who believe their own untruths and thus give off no tell-tale signs of deceit such as blushing or a racing pulse. And so, if any of several earlier hypotheses is correct, and concealed ovulation profits the concealer by deceiving others (whether these others be male or female), then perhaps self-concealment is just part of the package, advantageous because, as the wolf might have said to Red Riding Hood, "all the better to fool others, my dear."

Nevertheless, it is still difficult to understand why so important a fact about a woman's body shouldn't be shared with the woman herself, along with mechanisms that ensure that the information be shared with others only on a "need to know" basis. How is it that *Homo sapiens,* unique among living things in its self-awareness, has remained so ignorant of something so basic?

Biologist Nancy Burley puzzled over this question and came up with a novel answer: maybe modern-day women are kept in the dark about this fundamental aspect of their own bodies because in the past those who could tell when they were ovulating might have had *fewer* children. If so, then evolution would have favored those who didn't know when they were ovulating.[41]

We find this consciousness hypothesis especially interesting not only because it appears counterintuitive, but also because it brings in another factor not usually discussed by evolutionary scientists: human consciousness. Burley's brainstorm rests on one indisputable observation and one highly likely speculation. The observation: compared with animals, human beings have a hard time with pregnancy and childbirth. The speculation: even our prescientific, nontechnological ancestors may have realized early on that there was a connection between sexual intercourse and baby making. Add now a plausible connection between the observation and the speculation: because childbirth is painful and sometimes lethal, women who could tell when they were ovulating might have been especially inclined to avoid sex at those times, just as many women do today. If so, who would have done more than her share of the copulating and reproducing? Those whose ovulation was *least* apparent to themselves. While the woman whose ovulation was more chimpanzee-like or just more traditionally mammalian was taking advantage of her awareness of this information by having fewer children, the concealed ovulators were unwittingly inheriting the earth.

Burley examined the anthropological record. She found persuasive evidence that in most human groups, men are more enthusiastic about being fathers than women are about being mothers, especially when it comes to having large families. After all, without the option of anesthesia and Cesareans if necessary, the biological cost of childbirth is heavy and borne entirely by women. Until quite recently, even in the Western world, women often died giving birth, whereupon their husbands simply remarried. No doubt, death in childbirth was a family tragedy as well, but the hard fact remains that it is most tragic for the woman who dies! All this makes it possible that our earliest grandmothers who detected their "fertile time of the month" might well have made special efforts at those times to fend off the advances of our earliest grandfathers. We call this explanation the Headache Hypothesis.

Women who used their brains to keep one step ahead of their fertile bodies (and away from their presumably ardent male consorts) might well have succeeded, at least to some degree. But the result, ironically, would have sabotaged the system: Who would have gotten pregnant? Not those who could detect their own ovulation, but those who couldn't, those who were unaware of what was going on inside their own bodies. A case of matter over mind. If so, then by an ironic twist of biology, our subsequent ancestors would have been those who not only didn't reveal any cues as to whether they were ovulating, but themselves couldn't tell when they were doing so.

So there it is. Or rather, there *they* are: the many explanations for the evolutionary enigma known as concealed ovulation.

The abundance and diversity of these various scientific stories should engender celebration, not discouragement. After all, with Mark Twain's numerous attempts to stop smoking, each failure was just that: a failure. But in attempts to explain something so complex as concealed ovulation, the fact that there are many possible answers may also imply just that: perhaps there really *are* that many answers! After all, most of them are not mutually exclusive, so each plausible explanation is a possible success.

Most people look for *the* reason for something: why the car won't start, why a particular meal tasted so good, and so forth. The idea is that for every effect there is a cause—that is, one cause. But this needn't be true, especially when it comes to something as elaborate as human beings. With regard to concealed ovulation, there is no reason why biology couldn't have been responding to many different factors, several of them pushing in the same direction. Two or more good evolutionary reasons for something may be even better or more effective than one. (Of course, it still remains a challenge to figure out which factor—or factors—might have pushed especially hard and which might just be nifty testimony to human ingenuity rather than to biology.)

Considerable scientific attention is focused on the question of concealed ovulation not only as a matter of theoretical biology, but because of its practical implications for birth control. It is therefore not unlikely that within a few years the matter will be resolved, maybe by someone reading this book. For now, however, concealed ovulation is still an enigma.

Next up: breasts. Whereas the mystery of ovulation is why it is so secret, the mystery of breasts is exactly the opposite: Why are they so obvious? Whereas human ovulation is mysterious because something so important is so hidden, human breasts are mysterious because something so unimportant most of the time is so prominent most of the time and gets so much undeserved attention nearly all the time.

At first glance, the purpose of breasts seems as evident as they themselves are. Like all mammals, human infants require milk, and adult females are equipped to provide it. This combination should obviate any puzzlement over why breasts are widely seen as not only aesthetic, but also erotic—although to varying degrees in different societies—and not just as milk-delivery systems. Even in topless cultures, female breasts and nipples are typically considered to be sexually charged and are fondled during intercourse.[1] Insofar as sex appeal is related to indications of health, fertility, and the promise of evolutionary benefit conveyed by those perceived as attractive, there is nothing especially perplexing about the fact that women have breasts or that men like them. As we'll soon see, however, human breasts are also genuine biological enigmas.

4

_____

*Breasts and
Other Curves*

To be sure, all mammals have mammary glands; that's where the word *mammal* comes from. In his *Systema naturae* (1735), Linnaeus, the great taxonomist, initially used the name "Quadrupedia" for the biological class in which humans were placed. Many naturalists were outraged, however, that the human species was being characterized as an animal with four feet (as well as four incisors). Close rivals, for a time, were the names "Pilosa"—from the word for "hairy"—suggested by the naturalist John Ray and "Tetracoilia," referring to the four-chambered heart.[2] Even today, German biologists use the term *Säugetiere* (suckling animals), thereby focusing on the process—the act of giving and getting suck—rather than on the breasts and nipples being thus attended.

For reasons that aren't entirely clear, however, by the tenth edition of the *Systema* (1758), "Quadrupedia" was out, and our class became known as "Mammalia," literally meaning "of the breast." No other comparable animal group has been named for a physical trait of this sort and, moreover, one that is prominent in just half the population only some of the time. This designation also broke with centuries of sexist tradition that has made maleness the measure of all things. Although the breast may well deserve accolades as a masterpiece of female biology (certainly, its male counterpart is laughably meager by contrast), the term *mammal* doesn't necessarily rank as a nomenclatural, protofeminist triumph. Thus, a more pessimistic interpretation would note that biologists settled on a classically female trait, lactating mammary glands, to link us to certain other animals (identifying that thing that cats, rats, and human beings have in common), whereas when it came to identifying our species-specific uniqueness, Linnaeus turned to a characteristic long thought to be particularly male—namely, reason—thereby coming up with *Homo sapiens.*

Regardless of their sociological or historical significance, prominent nonlactating breasts are biological entities and puzzling ones at that. Human beings excepted, it is unheard of in the animal world for *non*lactating females to have well-developed mammaries. Why have women been blessed—or, in some people's opinion, cursed—with such prominent additions to their anatomy? The obvious explanation is that breasts are an unavoidable consequence of selection for ability to nourish offspring. However, nonlactating human breasts are composed almost entirely of fat, not milk-producing glandular tissue, which develops during the late stages of pregnancy.

Common sense would suggest that there is at least some correlation between low to average body mass index (BMI) and breast size, and between BMI and milk production within this range, but we haven't been able to locate any clear evidence of this.

In any event, unlike the case of menstruation, which at least some biologists see as an unavoidable consequence of something else (e.g., the need to remove pathogens, the cost of otherwise maintaining a metabolically expensive uterine lining, and so forth), no one to our knowledge has seriously proposed that the breasts of nubile, nonpregnant, nonlactating women must be well filled with adipose tissue (where fat is stored) in order to make milk later. Why not? Because no other mammal needs fatty breasts in order to lactate. The fat residing in the human breast isn't readily mobilized into milk, despite its geographic proximity; when times are tough, lactating women are more likely to reach (figuratively) into their hips, thighs, and arms for the nutrients needed to produce breast milk. Moreover, too much adipose tissue actually gets in the way of milk production.

All of this leads to the conclusion that breasts are clearly functional, but not merely so—in other words, they aren't just milk bottles. The most likely alternative explanation for their continued presence? Social signaling. And for whom are these signals intended? Almost certainly men.

"There they go again, how like men," indignant readers (of both sexes) might complain at this point, "to assume that breasts must be pointing at them." But in fact they probably are. When providing milk for an infant, breasts are doubtless doing their evolutionary thing. But even when they aren't serving such an obvious nutritional function and are simply there, whether fully exposed or peeping out demurely or alluringly from various coverings, then they are *still* doing their evolutionary thing. These two "things" are quite different, albeit ultimately connected by a strong thread of biological propriety. Breasts are breathtaking, bilateral, beautiful, bountiful (sometimes)—and baffling, at least to the evolutionary biologist. Breasts are also intensely practical as devices for nourishing infants, yet evolutionists must be excused for thinking that they must have evolved with men in mind, especially because the likelihood is that they have.

Breasts are amazingly diverse, perhaps more so than any other organ in the human body. They can be huge and pendulous, small and pert, spherical or cylindrical, nearly symmetrical or not very, sources of pride and of anxiety, generators of comfort, nourishment, attention, embarrassment, and lust. Something to hide or to flaunt. If nothing else, all this variety—in structure, not to mention function—suggests that breasts are not unidimensional in their biological role. Otherwise, if breasts served a particular, constrained evolutionary function, selection would likely have narrowed their anatomical range accordingly, even as their cultural elaboration has diversified.

The mechanism that links milk-making practicality and sex-evoking sensuality is, of course, natural selection. When most people think of natural selection, they focus on "survival of the fittest," a regrettable phrase invented by Herbert Spencer, not Darwin, and that has two particular disadvantages. First, it seems to be redundant: the fittest are those that survive. How do we know they are fittest? Because they survive. In fact, however, the term *fitness* in the Darwinian sense refers to reproductive success, so that the biologically meaningful take-home message is that genes that are more successful in projecting themselves into succeeding generations will "survive" and prosper over evolutionary time.

A second problem with "survival of the fittest" is that it places too much emphasis on survival. Not that survival doesn't matter: after all, most individuals—

genes—have to succeed in building an adequate body, one that func-
physiologically and behaviorally well in order to be "fit" in the evolutionary
. But natural selection doesn't reward just success in pumping blood, breath-
air, finding food, avoiding becoming food for someone else, migrating if you
must, scratching when you itch. All these activities—sometimes called "survival
selection" or "ecological selection"—are important; all contribute mightily to
natural selection. But also important are activities and traits that not only maintain one's body, but help create and maintain the bodies of one's offspring: building a nest if you're a bird, making milk if you're a mammal. And so we are back
to breasts.

As Darwin pointed out, there is another component to natural selection in
addition to its directly functional, ecological, survival, and baby-making side.
He called it "sexual selection," and although such selection is perfectly "natural,"
adhering to the same biological bottom line of reproductive success, it appears—
at least at first blush—to be a horse of a different color. Many traits appear disadvantageous if we take only a narrow ecological or survival perspective. The classic
case is the peacock's tail, a seeming case of biological excess if ever there was one.
A peacock doesn't need a six-foot-long, gaudily painted monstrosity in order to
survive. In fact, it takes a great deal of genetic capitol and metabolic energy to
make the darned thing, which ends up making the bird more conspicuous to
predators and can even get caught in the bushes. If you are a peacock, however,
you evidently must have that tail in order to mate. Hence, sexual selection.

A trait is sexually selected if it helps obtain a mate. The name of the game
is still natural selection, as natural as can be considering that the ultimate
adaptive significance of a sexually selected trait—no less than that of straightforward ecologically oriented selection—is a matter of evolutionary fitness.
Sexual selection is nonetheless worth singling out, however, because of the
paradoxical fact that in such cases evolution ends up favoring a characteristic
that on the surface should have been selected *against.* But if you could get
inside the head of a peahen, you, too, would find the peacock's tail appealing
rather than surprising, which helps explain why peacocks have them.

Back to breasts. Being conspicuous even when not making milk is a human
mammary specialty, such that even a flat-chested woman is buxom compared
to any other nonlactating mammal. As noted, human breasts have a very high
ratio of fat to glandular tissue; from an engineering standpoint, it does not
seem that milk production is the goal being optimized. What, then, is? Almost
certainly, female breast development, like the peacock's tail, is at least partly a
result of sexual selection.

Whenever biologists find a trait in one sex, regardless of species, that is of particular interest to members of the other sex, their scientific antennae start quivering to the tune of sexual selection. Thus, the fact that men find breasts especially interesting is sufficient to suggest that something more than the practicalities of milk making are at issue. Moreover, when the trait in question carries with it obvious practical *disadvantages*—such as a peacock's tail—sexual selection once again emerges as a prime suspect. And breasts, like a peacock's tail, are costly.

Any physically active woman will confirm that large breasts (or even small ones) can be troublesome—hence, the existence of athletic bras. Although, as we shall see, breasts may well be functional as nutrition-storage organs, it would have been a far better design to store extra calories on the hips, butt, or upper arms, where they can be wrapped around solid bone rather than left unsupported. Moreover, breasts are not only inconvenient, but downright dangerous because they are prone to cancer.

All things being equal, in the United States a woman's lifetime risk of breast cancer is 13 percent, with roughly two hundred thousand women diagnosed in 2007 alone. This vulnerability may be associated with the unusual growth potential of breast cells, and cancer results from uncontrolled cell growth. In order to remain normal, breast tissue probably requires more cellular restraints on growth than are needed by most other body organs because breasts change dramatically during a woman's normal life, expanding during lactation and then shrinking. Breast cancer is not genetically "caused," but it is statistically associated with defects in either of two genes—Breast Cancer gene 1 (BRCA1) and Breast Cancer gene 2 (BRCA2), which in their normal form produce a protein that inhibits excessive growth of glandular tissue. Also involved in the development of breast cancer is a tumor suppressor gene known as "deleted in breast cancer" (DBC), which, when rendered nonfunctional by mutation, results in tumor development. Adding to the complexity is the fact that "breast cancer genes" are responsible overall for a very small proportion of breast cancer cases; there evidently is a huge environmental component. Moreover, some men are also victimized by breast cancer. There is no question that breast cancer, whatever its cause—or, more accurately, its causes—is a huge problem, and one that is closely associated with having breasts.

Prominent and permanent mammary glands, in short, aren't simple or without cost.

To these factors must be added the immense cultural overlay that breasts bear. "A woman's breasts," writes Natalie Angier, "welcome illusion and the imaginative opportunities of clothing. They can be enhanced or muted, as a woman chooses, and their very substance suggests as much: they are soft and flexible, clay to play with. They are funny things, really, and we should learn to laugh at them, which may be easier to do if we first take them seriously."[3]

And seriously is how we shall take them. After all, the subject is intriguing to scientists because the breast, more than any other body part, does double duty: as an ecologically selected organ (for nourishing infants) and as a sexually selected one (for attracting and/or retaining mates). Exactly how and why it balances these two duties remains an evolutionary enigma and the subject of this chapter.

A Buttocks Mimic?

Let's first describe—and, we hope, dispense with—the most notorious attempt to explain breasts' dual function: the buttocks mimic hypothesis. In his best-selling book *The Naked Ape,* British ethologist Desmond Morris suggested decades ago that human breasts are visual reminders of the buttocks, transplanted (or, rather, expanded) on the female chest in order to encourage the male to mate ventrally (i.e., facing his partner).[4] The idea may seem downright foolish, but in fact it continues to be taken seriously today—or at least to receive deferential mention in most treatments of the evolution of human sexuality in general and of breasts in particular.

Even though the buttocks mimic hypothesis itself has few if any current proponents, it captured its share (perhaps more than its share!) of scientific attention because, despite seeming so improbable, it emphasized an important aspect of human biology: the role of social bonding and biparental care. These evolutionary developments are widely considered crucial for the human species because human infants are born helpless and in need of considerable protection and assistance. Accordingly, it seems reasonable that *Homo sapiens* employ male-female bonding as a way of guaranteeing care for their babies, and such care, in turn, is most likely when there are two adults involved rather than one. This final observation, at least, is not in serious dispute (we already encountered it in several forms when seeking to explain concealed ovulation). Morris's initial claim, however, needs to be critically examined.

The suggestion is that because most mammals mate dorsoventrally, males have been selected to respond to the females' rounded buttocks. But dorsoventral mating, he suggests, isn't very personal, and it is precisely the person-to-person component of sexual intercourse that is needed to cement male-female bonding. Ventroventral mating, by contrast, is more likely to be—or at least capable of being—eye to eye and therefore soul to soul. Couples who achieved such a connection would presumably have been better bonded and therefore more likely to share in childrearing; their children would not only have been more successful, but also more inclined to mate ventroventrally in their turn because they carried with them their parents' genetically influenced propensity for face-to-face mating. Ipso facto, selection would have favored women whose breast development inspired their mates to copulate frontally.

It would be nice if such a remarkable and novel idea could be embraced. Frontally. Nicer yet if it could be tested, but as with so many such hypotheses, postulating is easier than evaluating. And in this case at least, the preliminary evaluations are discouraging. In theory, it should be possible to determine whether there is any correlation between a woman's breast development and her partner's preference for sexual position. We are not aware of any studies to this effect and would be very surprised if the prediction were confirmed. In all fairness, however, we can only note that the jury is out. (The buttocks mimic hypothesis doesn't make any predictions about women's predilections.)

Logic and the available evidence offer nothing but bad news for the buttocks mimic hypothesis. Although quadrupedal mammals nearly always mate dorsoventrally (hence the phrase "doggy style"), not all primates do. Gorillas, orangutans, and bonobos use a variety of positions, including ventroventral, but their females are nonetheless flat-chested. And although strong monogamous pair bonds are generally quite rare among mammals, they do exist in a few species, including the occasional antelope, some foxes, beaver, otters, California mice, pygmy marmosets, as well as the fat-tailed dwarf lemur and the Malagasy giant jumping rat. How do these creatures copulate? Dorsoventrally. Therefore, prominent breasts are not necessary for frontal mating, and frontal mating is not necessary for pair bonding.

It is, in addition, a wide stretch indeed to imagine that male mammals of any species need the visual image of female buttocks to stimulate them to mate. Pheromones and behavior usually do the job quite nicely. Moreover, mammalian buttocks are covered with hair and do not present themselves as naked, rounded globes. It seems equally probable—which is to say, quite improbable—that natural selection worked in the other direction so that human beings' rounded buttocks evolved to mimic breasts! Even Just-So Stories, much as we adore them, must be monitored for plausibility. Sometimes the imagination needs weights rather than wings, and the buttocks mimic hypothesis not only is a case in point, but lends itself to precisely the kind of extravagant over-imagining that must be guarded against (but is, on occasion, fun to indulge). For example, if buttocks evolved to resemble breasts rather than vice versa, what about the possibility that paired, breastlike buttocks encouraged ancestral toddlers, not yet weaned, to approach their mothers whenever the latter happened to be facing away? Sure enough, buttocks are appropriately low on the body, at the perfect height to maximize their visibility to prodigal two-year-olds. This "hypothesis," moreover, might explain why human buttocks aren't located on the shoulders!

We're describing improbable—indeed, ridiculous—notions in the hope of immunizing the reader against *hypothetica ad absurdum,* so here's another one, which, like the buttocks mimic hypothesis, warrants attention if only because of the undeserved attention it has already received. Writer Elaine Morgan has long championed the bizarre notion that people evolved as "aquatic apes," among whom, she claims, breasts evolved as flotation devices.[5] After all, sailors during World War II used to refer to their life vests as "Mae Wests." Babies, according to Morgan, could cling to their mothers' breasts as to water wings. But what about men? Well, presumably they also would have clung to those breasts. After all, they do so whenever they can. A more serious response is that if breasts were adaptive for our aquatic ancestors because of their literal buoyancy, then men and even children should have evolved them, too, rather than relying on adult women to keep them afloat.

Human buttocks almost certainly achieved their current configuration because of various mechanical considerations associated with walking (not swimming) upright, and human breasts likewise evolved to their current specifications because of altogether separate reasons. What might they be?

Let's get serious.

Of Calories, Fat, and Deception

To start, we feel obliged to raise one possibility that is more parochial than alluring, but still worth considering: the fatty deposits that make up the largest part of a nonlactating breast might simply have evolved as a way of storing calories during our long Pleistocene adolescence, during which human ancestors lacked refrigeration. Thus, breasts could have been a kind of natural, built-in food pantry whose contents didn't spoil. If so, then their initial payoff was simply a function of survival selection, advantageous whether men noticed them or not. No one should be surprised that men would then be attracted to such a larder, once it was initiated, and in proportion to how well it was stocked.

There is little doubt that fat in sufficient quantities is crucially important for successful pregnancy and lactation. Among bears, for example, embryos don't implant unless the mother-to-be possesses a minimum amount of body fat, enough to ensure a good chance of success in supporting a pregnancy and lactation. So perhaps protruding breasts among humans are a signal of "adequate adiposity." After all, we are bearlike in the connection between fat reserves and successful reproduction: menarche is delayed in adolescents who suffer from anorexia, and it isn't uncommon for highly trained aerobic athletes to report that their periods have ceased altogether. Pregnancy itself requires a woman to expend approximately 75,000 calories beyond her normal maintenance needs, after which the energetic costs of reproduction are just beginning: lactation takes an additional 650 calories per day or nearly 20,000 per month.[6]

In other words, every four months of nursing exceeds the caloric needs of an entire pregnancy, so that two years of nursing (which is probably less than the pretechnological average in the millennia before bottle feeding) equals six times the already substantial demands of pregnancy. In addition, body fat represents a kind of famine insurance, especially important for a creature whose food availability may have fluctuated unpredictably. Our ancient ancestral mothers—those who succeeded in carrying their babies to term and then in nourishing them—must have been closer to Earth Mother goddesses than to anorexic fashion models, and their shape, because they were metabolically functional, likely did not go unnoticed by their prospective sexual partners.

What people find compelling or romantically appealing seems likely to correspond to what ultimately led to reproductive success, just as berries are more gastronomically compelling than bark and a healthy smile is more appealing than a misery-laden frown. Deeply imprinted on the human mind is doubtless a formula something like "contributes to reproductive success" = desirable to mate with," just as "good for you" = "desirable to eat," which explains why good food is generally appealing to people, just as good health and a benevolent disposition are appealing to *Homo sapiens*' romantic taste buds as well.

"No one man in a billion, when taking his dinner," wrote William James, "ever thinks of utility. He eats because the food tastes good and makes him want more. If you ask him why he should want to eat more of what tastes like that, instead of revering you as a philosopher he will probably laugh at you for a fool."[7]

Substitute "evolutionary fitness" for "utility" and "biologist" for "philosopher," and William James might have been describing how evolution operates to hone human food preferences. Hungry people don't generally associate their internal sensation with a deprivation of carbohydrates, amino acids, or lipids, any more than they are motivated by awareness of the details of cellular metabolism. In fact, however, James's "utility"—which is to say, Darwin's "fitness"—underlies such preferences. Selection has almost certainly generated human sexual preferences in response to the preferences' fitness consequences. Just as a person who eats fruits, vegetables, grains, and meat would be favored by evolution over someone who prefers to eat dirt, stones, and wood, a person who prefers to form a relationship with someone who is healthy, wealthy, generous, and wise, and thus likely to contribute to the chooser's success, will be favored by evolution over someone whose preferences are less likely to be reproductively rewarded.

By the same token, if abundant breast tissue signals adequate calorie storage, selection might have favored (for women) concentrating fat in such a conspicuous location and (for men) responding positively to it.

The result is a possible explanation of why female fat is so often stored where it is: in the service of both calorie-driven practicality (fat storage itself, for its own sake) and in the promise of lactation (that may or may not be fulfilled). Normal women concentrate their fat in breasts and hips, a tendency that may owe its existence to false advertising: the former indicating lactational capacity and the latter a wide birth canal. Note that wide hips—if produced by fat—don't say anything about width of the birth canal, just as pre-pregnancy breast size isn't correlated with ability to lactate. Both, then, may well be deceptive, albeit attractive to men, who, by contrast, store fat especially in their waists because they needn't promise competence in either breast-feeding or in giving birth.

As mere storage organs, breasts are inconvenient—especially when large—and probably wouldn't have evolved if storage were their only function because, as we have seen, it would be more efficient to accumulate calories on the hips and buttocks. But in fact their signaling role may be equally, if not more significant. The resulting deception hypothesis suggests a degree of false advertising, tinged with strategies and counterstrategies, deception and truth telling, ending up not only with female breasts but also with the seemingly universal male preference for "shapely" women.

The deception hypothesis begins with the fact that breasts indeed enlarge while a woman is nursing, such that the connection between a ripe, full bosom and maternal nutrition-giving hardly requires a massive intellect; thus, it is a connection that even Stone Age men (and we do mean *men*) were able to make. It's the most obvious breast-explaining hypothesis of all: that breasts exist to provide nourishment and that they are attractive to men for the same reason. In James Joyce's *Portrait of the Artist as a Young Man,* Stephen Dedalus muses with his friends on the nature of female beauty, whereupon he concludes that "every physical quality admired by men in women is in direct connection with the manifold functions of women for the propagation of the species. . . . You admired the great flanks of Venus because you felt that she would bear you burly offspring and admired her great breasts because you felt that she would give good milk to her children and yours."[8]

If so, then as we have seen, such admiration is based on a likely misunderstanding, albeit a creative one that may have opened the doors of deception whereby our female ancestors induced their male admirers to (mis)direct some of their admiration to organs stuffed with fat—which is cheaper to maintain than active glandular tissue—rather than to the real thing (organs full of milk).

Some body organs, probably the majority of them, are honest. There is no reason for kidneys to lie about their capabilities, if only because they cannot be perceived by a prospective admirer. By the same token, muscles are basically truthful: if they are large and bulging, you can count on them being strong. But breasts, for all their appeal, aren't guaranteed to be good at making milk. In fact, just as excessive muscle development can cause someone to be "muscle bound" and thus less functional than meets the eye, excessive mammary development can make a woman "breast bound" and less capable of lactating than her smaller-bosomed sisters, whose breasts expand when the need arises. In short, prominent nonlactating breasts may have evolved insofar as women succeeded in promising something that they may not have been able to deliver, and men were deceived (albeit eagerly).[9]

If so, it can also be argued that men were not entirely victimized and helpless. They developed a kind of counterstrategy: so as to avoid being entirely fooled by the simple substitution of fat for gland, men are typically aroused by an "hourglass" female figure rather than by a spreading waist, which would indicate body fat in general. Consistent with this interpretation, evolutionary psychologist Devendra Singh has discovered that men prefer a waist-to-hip ratio (WHR) of 0.70, and that this preference appears to be a "cross-cultural universal," true in Bangladesh and Brazil no less than in Brooklyn.[10] Given a choice, men say they prefer a WHR of 0.8 over 0.9, and 0.7 over 0.8 (the higher the WHR, the larger the waist circumference compared to hips).

It is worth noting that the WHRs of prepubescent boys and girls are indistinguishable. Then they diverge, with girls depositing fat on their hips and upper thighs, whereas boys lose fat from their thighs and buttocks. Healthy adult women have a remarkable 40 percent more fat in their lower trunk than do men and WHRs that average between 0.67 and 0.80; the WHR for healthy men, by contrast, is between 0.85 and 0.95. If these differences derived from sexual selection, then after menopause women's WHRs should approach that of men, and men's should remain more or less unchanged throughout their adult lives, which is indeed the case.

Such findings, especially the oft-confirmed male preference for women with a low WHR, has given rise to an avalanche of research, much of it quite recent and nearly all consistent with the deception hypothesis.[11] However—and regrettably, for anyone seeking closure on this matter—these findings do not rigorously exclude other explanations for the evolution of breasts. The problem is that men's preference for "shapely" women does not necessarily explain why women are shaped as they are. Maybe men simply prefer women as they are (regardless of how they came by their characteristic anatomy), rather than their preference's having shaped the shapeliness in question. Moreover, although evolutionary biologists are increasingly agreed that the evolution of human breasts owes as much to signaling as it does to lactation, there is no agreement as to what specifically is being signaled.

Provisioning?

As to breasts' possible signals, one of the most interesting is also one of the most obvious, perhaps so apparent that to our knowledge it hasn't yet been suggested. Here goes. Breasts expand during lactation in human beings as in other mammals. Therefore, an especially full breast is likely to be a lactating one. Because lactation is energetically expensive, and successful nursing benefits the parents no less than the infant being fed, it seems reasonable that prehistoric men may have been especially predisposed to protect and provision prehistoric women who had enlarged breasts. Women with less-developed breasts—those closer to the mammalian, primate norm—would then have attracted less male attention and been less successful. The result would have been the evolution of women with comparatively large breasts, essentially lactation mimics.

Not only is lactation demanding, but lactating women, at least in nontechnological societies, are generally less able to obtain calories because they are

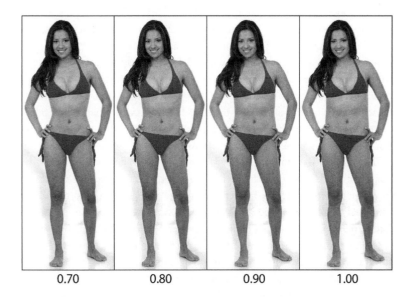

Photographically modified images of the same person with four different waist-hip ratios. Photograph courtesy of Jaime Confer.

encumbered with an infant and perhaps have less time available for foraging and gathering. At least one study found that among a group of hunter-gatherers—the Aché of Paraguay—lactating women obtain substantially fewer calories per hour of effort than do nonlactating women.[12] They can use a hand, and men—as everyone knows—are generally more than willing to provide it.

Ethologists talk about "releasers" among animals, or conspicuous traits such as the bright feathers of certain birds, that evoke various automatic responses from members of the same species. For example, male mountain bluebirds aggressively repel other males that approach their nests; they will also attack with equal vigor a small rubber ball that doesn't look at all (to a human observer) like a bird, but is blue like themselves. In such cases, cognition is not involved—the response is preprogrammed and "instinctive," automatically "released" by the presence of a key stimulus. Biologists have also found that if they artificially exaggerate the key characteristic of a releaser, making it "supernormal," the response it elicits is comparably over the top. For example, consider the American oystercatcher, a shorebird about the size of a crow. These animals, whose eggs are somewhat smaller than those of a barnyard chicken, will preferentially perch upon a watermelon if it is painted with the appropriate speckled pattern. According to the provisioning hypothesis, enlarged breasts are attractive to men as indicators of successful lactation, promising to yield a good reproductive return on their parental investment. If so, then in this case maybe the more the better, even if the "promise" is illusory, just as the American oystercatcher prefers an oversized "egg" to the real thing. What are now normal female breasts may therefore have evolved because of a male penchant for the "supernormal," even if it was somewhat disconnected from reality.

This supposition, in turn, leads to an interesting and as yet unresolved debate among evolutionary biologists, one that goes beyond the specifics of bluebirds, oystercatchers, or human breasts: How disconnected from reality is natural selection likely to get? Oystercatchers who incubated watermelons instead of their own eggs would certainly have left fewer descendants than would those whose preferences were more realistic and who stuck to their genuine reproductive interests. Their susceptibility to supernormal releasers is revealed only when researchers step in and generate an artificial situation not normally experienced. During the evolution of oystercatchers, therefore, natural selection was in a sense free to endow these animals with a simple algorithm— bigger is better—in part because such absurd exaggerations as watermelon-size eggs weren't normally encountered.

During the typical course of evolution, it is likely that neither traits nor preferences can stray too far from reality, especially if those traits and preferences are regularly tested for their impact on fitness. Accordingly, if oversized breasts seriously interfered with lactation (as well as with male preference for them), they would have been selected against. But if breasts somewhat larger than the primate norm enabled ancestral women to attract better mates, and if being attracted to such women enabled ancestral men to achieve enhanced reproductive success, then evolution may have favored both the trait and the preference despite the fact that by strict ecological/survival criteria, both might be less than optimal.

According to the provisioning hypothesis, men who were especially attentive to women who were nursing (and hence were large-breasted) may well have shared an additional evolutionary benefit with the object of their attentions: shortening the "interbirth interval" and thereby increasing their own reproductive success as well as that of their breeding partners. An anthropological study in Gambia found that on average 19 percent of conceptions occurred within eighteen months of the birth of a woman's previous child, whereas when nursing mothers were provided with additional food, significantly more conceptions, 33 percent, took place within a comparable timeframe.[13] If being provided extra food during nursing enhanced the reproductive prospects both of ancestral women and of the men who provisioned them, it would not be surprising that men evolved to respond positively to swollen breasts and that women thereby endowed experienced an evolutionary advantage. The result? A selective boost to women whose breasts generated enhanced male provisioning and to men with such a preference.

The provisioning hypothesis must be seen as, well, provisional, although evidence from nonhuman primates is somewhat supportive. On the downside, provisioning of females is relatively rare among primates in particular, and in no cases have males been seen to solicitously provide extra food to a pregnant or nursing female. Yet male chimpanzees are known to use food—especially meat—as a means of gaining sexual access to females.[14] It need not have been simply a matter of male provisioning's leading to enhanced reproductive success of their offspring, because female behavior may have conveyed greater fitness upon males who also used food as a breeding tactic, regardless of whether they bestowed it preferentially on their own lactating mates.

Being provisioned would also have been in the female's interest throughout her cycle, not only during pregnancy and lactation. This characteristic, in turn, may have selected for women who maintained enlarged breasts independent of their reproductive state. Pleistocene era women who already had enough calories on board to expand their breasts via fat deposition would have been the most likely to stimulate males to provide yet more nutrition. And so an early hominid female in a position to benefit from storing fat somewhere on her body might as well have done so via her breasts in order to stimulate additional male investment. This relationship is reminiscent of what has been called the "banker's paradox," so named because banks are least inclined to give money to prospective borrowers who especially need the money and who are therefore poor credit risks, while eagerly bestowing funds on those who need it least. In short, those who have and don't need much more, get! Something along these lines may have induced evolution to exaggerate early female hominids' breasts because those thereby endowed would have profited from men's increased self-interested largesse due to their inclination to "lend" resources to a prospective mate deemed to be a good investment.

Incidentally, just as no research attention has been paid to the Provisioning Hypothesis and little to signal-related sexual selection in general, there has essentially been none to their flip side—that is, the preferences of the sex being selected. For our purposes, we have to ask, Are there correlations between a woman's desirability and her preferred traits in a man? One basic prediction is that women with a low WHR—those women generally acknowledged to be more sexually desirable—would be more demanding when it comes to male resources. Anyone not surprised to see images of luscious young women being escorted by wealthy, often middle-aged or elderly men would not be surprised at this prediction, which is supported by the admittedly skimpy evidence available thus far: women with lower WHRs appear to be sexually choosier; in particular, they want wealthier men.[15]

Beyond the possibility that breasts evolved because they encouraged provisioning of women by men, there is the related notion that breasts' structure evolved to facilitate the provisioning of infants by their mothers. According to this seemingly commonsense view, first expressed by two anthropologists more than thirty years ago, the important thing about human breasts isn't that they are large, but that that they hang down, which "lowers the nipple height, and increases its flexibility, allowing the human infant to nurse more easily."[16] The authors went on to emphasize that because nursing babies are unable to cling on their own (and, moreover, there is little for them to cling to because women lost their body hair long ago), a nursing baby had to be held in the mother's lap or on her hip. As a result, infants couldn't reach a nipple that didn't dangle. The human breast could thus be seen as a purely functional, practical response to the evolution of bipedalism.

The problem with this practical provisioning hypothesis, however, is the same as the problem with explaining the evolution of breasts in general. There is no reason why human breasts couldn't have expanded during lactation, which would have placed them within reach of a baby, but then shrink, as is the case with all other mammals. Moreover, if the key to the evolution of human breasts is the extent to which they droop, then why doesn't droop develop with the first pregnancy rather than being so clearly correlated with age? Moreover, why don't men find drooping breasts especially appealing? And why do young, nubile women develop protruding breasts *prior* to becoming sexually mature, when there is no practical provisioning payoff?

Enter: Engineering and Honesty

Deception may attract our attention, eliciting as it does a frisson of cloak-and-dagger intrigue, not to mention the ethical alarm bells that go off simply because no one wants to be deceived or even to perceive herself as a deceiver. But it may turn out that honesty is the best policy, not only ethically, but evolutionarily, too.[17]

If there is any validity to the provisioning hypothesis, such that men were at least initially predisposed toward women whose bodies mimicked lactation, then men would also have been at risk of being duped by women whose large-breastedness resulted from pregnancy and who might therefore have been carrying another man's child. In this case, men should have evolved to find large breasts unappealing, even downright repulsive. One way for men to avoid this fitness downside would have been to insist that even as a woman's breasts were relatively full, her waistline wasn't. (As pregnancy advances, of course, WHR rises, eventually exceeding not merely 0.70, but 1.0.)

Even when a woman is not pregnant, the characteristically female pattern of fat deposition is proximately caused by high levels of estrogen (more accurately, high ratios of estrogen to testosterone), which promotes fatty hips—and breasts—and inhibits a fat tummy. A low WHR is therefore an honest indicator of a fertility-friendly hormonal balance (high ratio of estrogens to androgens) and in turn of reproductive potential.[18] The optimum female form should therefore possess an ample bosom and a low WHR, a combination that women probably cannot easily fake and that men, not surprisingly, employ as an assessment device (albeit unconsciously). Women would therefore be especially attractive if they have not only discernible breasts, but *at the same time* a low WHR. This seems to be the case not just today, but far back in time. Speaking of the Goddess Durga, the ancient Hindu epic *Bhagavata Purana* (sixth century c.e.) notes that "[b]y the magic of powers she assumed the form of a beautiful woman . . . her hips and breast were full, the waist slender." There is nothing culture bound about admiration for an "hourglass figure."[19]

An emerging discipline sometimes called "neuroaesthetics" looks for the proximate brain mechanisms underpinning perceptions of beauty.[20] Among these neuroaestheticians, V. S. Ramachandran in particular has proposed a list of "eight laws of artistic experience" (later expanded to ten) constituting "a set of heuristics that artists either consciously or unconsciously deploy to optimally titillate the visual areas of the brain." In an oft-cited and influential paper, Ramachandran and coauthor William Hirstein write that

one of these principles is a psychological phenomenon called the peak shift effect: If a rat is rewarded for discriminating a rectangle from a square, it will respond even more vigorously to a rectangle that is longer and skinnier that the prototype.

We suggest that this principle explains not only caricatures, but many other aspects of art. Example: An evocative sketch of a female nude may be one which selectively accentuates those feminine form-attributes that allow one to discriminate it from a male figure; a Boucher, a Van Gogh, or a Monet may be a caricature in "colour space" rather than form space.[21]

Recall those bizarre supernormal releasers among bluebirds and oyster-catchers. It is especially exciting that work such as Ramachandran's has begun to elucidate the precise brain pathways that are activated in such cases. Of course, just as there is attraction to certain images (and presumably to certain sounds, odors, touches, and so on), there can also be repulsion from others. A magnetic resonance imaging study, for example, showed that "beautiful" images evoke activation in the emotional, limbic centers of the brain and stimulation in the prefrontal cortex, whereas "ugly" ones generate activity in the motor cortex.[22] Why the latter? The researchers speculate that it might be part of "preparation for action" in response to aversive stimuli, although it isn't clear why action—albeit of a different sort—shouldn't also be associated with positive, beautiful, attractive stimuli, such as women with a low WHR.

In a lecture presented for the BBC, Ramachandran in fact pointed to the deep, intuitive appeal of exaggerated WHRs found in ancient Hindu art, although without employing that term. After asking, "Are there such things as artistic universals?" the neurologist spoke as a biologist rather than as a philosopher or art historian:

Let's assume that 90% of the variance you see in art is driven by cultural diversity or—more cynically—by just the auctioneer's hammer, and only 10% by universal laws that are common to all brains. The culturally driven 90% is what most people already study—it's called art history. As a scientist what I am interested in is the 10% that is universal—not in the endless variations imposed by cultures. . . . For example if you go to Southern India, you look at the famous Chola bronze of the goddess Parvati dating back to the 12th century. For Indian eyes, she is supposed to represent the very epitome of feminine sensuality, grace, poise, dignity, everything that's good about being a woman. And she's of course also very voluptuous. But the Victorian Englishmen who first encountered these sculptures were appalled by Parvati, partly because they were prudish, but partly also just because of just plain ignorance. They complained that the breasts were way too big, the hips were too big and the waist was too narrow. It didn't look anything like a real woman—it wasn't realistic.[23]

But, of course, realism wasn't the point, just as it isn't the point in much great art, which flourishes at least in part by "releasing" certain built-in human preferences. Parvati's exaggerated shape is indeed supernormal, as befits a deity claimed to be supernatural. She also exemplifies the powerful, all-too-human preference for a low WHR, which in turn seems to have helped sculpt the altogether natural human body as we know it today.

A study published in the prestigious British medical journal *Lancet* examined the measurements of three hundred fashion models, three hundred "glamour" models (from *Playboy* magazine), three hundred normal healthy women, thirty bulimic women, and thirty anorexic women. It found that anorexic and bulimic women had the highest WHR (0.76 and 0.77), followed by normal women (0.74), then fashion models (0.71), and finally glamour models, whose WHR was an exaggerated, Barbie-ish, and nearly inhuman 0.68. Waist-to-bust ratios were, not surprisingly, arranged the other way: glamour models had the largest busts relative to their waists, and normal women had the smallest. Bust to hip ratios were almost exactly one to one for glamour and fashion models, whereas normal women averaged a ratio of 0.92—that is, their hips were significantly larger than their bustlines.[24]

No question, cultural priorities and expectations figure importantly as well, with plumpness generally admired in societies characterized by resource scarcity, not to mention the possible vagaries of local cultural tradition. It has been found, for example, that men in two different isolated Andean populations prefer "tubular" women.[25] Although preference for overall thinness has increased along with affluence, Devendra Singh found, for example, when examining images of beauty contest winners and *Playboy* centerfolds, that the preferred WHR has remained remarkably unchanged, at 0.70. This ratio is also a better predictor of female attractiveness than is overall fatness or thinness.[26] Even when the societal norm favors being generally well upholstered and plump (à la Rubens' models) or slender (current Western fashion models), the idealized relationship between waist and hips remains quite consistent.

And let's face it, what men want is nearly identical to what women want to be. This correlation might simply be a coincidence, but we doubt it. A more likely possibility is that both male and female preferences have converged independently on a female WHR that is biologically optimum for reproduction. Or perhaps one sex's preference has driven the other's.

By the way, these findings do not in themselves argue against the role of fat deposition as an honest indicator of female reproductive capacity because fat, when concentrated in breasts and buttocks, is less ambiguous than if spread

Statue of the Indian goddess Parvati, likely from the first quarter of the tenth century. Image from the Metropolitan Museum of Art, bequest of Cora Tinken Burnett, 1956 (57.51.3). Image copyright © Metropolitan Museum of Art.

EVOLUTIONARY HYPOTHESES EXPLAINING WOMEN'S BREASTS

Buttocks mimic: promoting ventral mating
Flotation devices: aquatic apes
Calorie storage 1: food pantry
Calorie storage 2: signaling resources
Deception: promising lactation
Provisioning: motherhood mimics
Practical provisioning: easier nursing
Engineering: counterweights
Honest signaling 1: releasers
Honest signaling 2: symmetry assessment
Honest signaling 3: nubility (Goldilocks) assessment
Honest signaling 4: to other women
Doughty daughters 1: selection for desirable daughters
Doughty daughters 2: selection for a handicap
Fine-feathered females: vying for paternal investment

over the entire body. We haven't said much so far about fat deposition below the waist, nor do we intend to. But a purely practical, engineering-based assessment of women's anatomy might suggest that sexual selection and male preferences may be overrated as explanations, at least in some cases. Protruding dorsal buttocks might, for example, simply be an appropriate counterweight, structurally adaptive for an upright biped that—for whatever reason(s)—has evolved to carry substantial mass on her ventral surface, not least during pregnancy. At the same time, the engineering hypothesis is weakened by the fact of substantial racial variation in the degree to which buttocks protrude (such as the famous case of the Hottentots), although no one as yet has satisfactorily explained how such variation might satisfactorily be linked to men's sexual preferences or to any other fitness-enhancing consideration.

In summary, male preference for a female bosom might (and we emphasize *might*) have been initiated by male inclination to provision women—especially those with whom they have been sexually involved—who suggest lactation and thus the likelihood of profiting from such attention and of benefiting the males who are attracted to them. If so, then women whose breasts were naturally augmented would have been more successful reproductively, as would males who chose them, which in turn may have selected for women whose breast size "dishonestly" mimicked that condition, until the trait—deceptive or not— became the norm.

Alternatively, breasts may serve as an entirely different kind of signal, an honest one indicating physical health and genetic quality. It is well established—although only recently, through the work of biologist Randy Thornhill and others—that people are likely to equate body symmetry with physical attractiveness; facial lopsidedness in particular is distinctly unappealing. As with research on WHR, the connection between anatomic symmetry and sociosexual preference has led to a remarkable recent outpouring of empirical validation.[27] (It is also noteworthy—and appropriate to our earlier consideration of concealed ovulation and the dispute about whether women experience estrus—that ovulating women show an enhanced preference for symmetric male faces just as men prefer symmetric women.)[28]

Starfish and sea urchins are radially symmetric; vertebrates are bilaterally symmetric. Although we take such symmetry—a close balance between right and left—pretty much for granted, it isn't easy to achieve mirror-image identity or even similarity. When given choices about human bodies and faces, people invariably prefer the more symmetric. Moreover, just as the preferred WHR of 0.70 correlates with reproductively optimal estrogen concentrations, body symmetry correlates with low levels of pathogens and few genetic abnormalities. Thus, the human preference for symmetry, not surprisingly, serves the preferer's biological interest. Translated into evolutionary terms, this statement means that individuals who preferentially mated with symmetric partners were more likely to be allying their genes with "good" ones, which in turn would promote the success of any underlying symmetry-preferring genetic factors.

Having noted this, we must also acknowledge that symmetry can explain only a small part of perceived human attractiveness. A truly ugly person, whose ugliness is utterly symmetric, left to right, would doubtless still be ugly! And some extraordinarily beautiful women, such as Marilyn Monroe, exaggerated certain facial *a*symmetries such as moles (revealingly called "beauty marks"). Symmetry nonetheless counts for something, especially if it is dramatically lacking, and because breasts are so prominently displayed left to right, they readily lend themselves to assessment—whether conscious or unconscious—in this regard.[29] Symmetry, in turn, provides information

Marilyn Monroe, showing conspicuous (and asymmetrical) "beauty spot." Photograph reprinted by permission of CMG Worldwide.

concerning the health and thus the reproductive prospects of the woman in question. Recent studies have interestingly lent support to this conjecture: women with symmetrical breasts have greater fertility than do their less-balanced counterparts.[30] Accordingly, early men who preferred women with symmetrical breasts would likely have experienced an evolutionary benefit as a result.

An interesting wrinkle here is that larger breasts are more likely to be asymmetric than are smaller ones,[31] a finding that is difficult to interpret. Is it because whatever causes enhanced asymmetry also generates increased size? Or might it be that greater size is simply more likely to reveal any underlying differences? Either way, just as men frequently worry about penis size, women often obsess about whether their breasts are suitably symmetrical. Indeed, one typically unmentioned but nonetheless genuine cosmetic payoff of brassieres is that they obscure underlying breast asymmetries.

Why have women gone along with being judged and evaluated in this way? Wouldn't they be better off obscuring any imperfections rather than presenting them conspicuously and literally right up front for everyone to see? For one thing, if attractively symmetrical women displayed their charms, and men insisted thereafter on this opportunity when it came to making a choice, then other women may have had little alternative but to go along, at least if they wanted a chance with the more desirable men.

We should probably also look deep into our evolutionary past because for obscure reasons—probably related to unavoidable developmental constraints—mammalian nipples and breasts have always been horizontally paired, left–right, and never vertically oriented. It would seem hard to prevent others from assessing the symmetry of such prominently displayed and obviously paired structures, whether they evolved to facilitate the evaluation or not. (If humans, like other primates, were covered with hair, breasts wouldn't lend themselves to being so readily judged, but for other reasons—which are also currently under debate—ancestral *Homo sapiens* evolved hairlessness.) Natural selection may have enlarged the size of the organs under scrutiny to mitigate the discrimination against women with asymmetrical breasts. An asymmetry between left and right breasts occupying, say, 375 and 400 cubic centimeters, respectively, would be less noticeable than if the volumes were 75 and 100 cubic centimeters. If so, then women's breast evolution would have involved making it harder for men to assess breasts by paradoxically making them *more* prominent.

The Goldilocks Hypothesis

More squarely in the realm of honest signaling, another possibility also deserves mention. Full disclosure: we thought we had originated the next hypothesis and that it would receive its first-ever exposition here and now, but an anonymous reviewer informed us that it has already been suggested by anthropologist Frank Marlowe, who called it the nubility hypothesis.[32] In short, what Marlow proposed is that perhaps breasts are an accurate signal—to men—of women's "residual reproductive value" (RRV). First elaborated by famed evolutionary theorist R. A. Fisher,[33] RRV refers to an individual's expected future breeding potential; it can be expected that males would be selected to prefer as sexual partners those females whose RRV is as high as possible—an expectation that applies especially to human beings, which frequently make long-term mating and child-rearing commitments.

If, by contrast, the interaction between male and female were only short term, males can be predicted to prefer females at peak fertility. This is precisely the case for our closest relatives, the chimpanzees, among whom older females are more likely to reward a copulation with a conception and males accordingly prefer older females to younger.[34] But if long-term bonding is at issue, then males should prefer females with maximum RRV—that is, not simply someone so fertile that she is likely to convert a given copulation into a maximally successful offspring, but rather someone who is just entering maturity and has the largest possible reproductive future ahead of her. Too young and she isn't yet ovulating; too old and she isn't ovulating anymore. The importance of assessing nubility is all the greater in a species that experiences menopause, which sets an upper age limit on a woman's possible reproduction (more on menopause in chapter 6).

For now, it should be clear that evolution likely rewarded men who preferred sexual engagement with women who were neither too young (prepubescent) nor too old (postmenopausal), but "just right." But because women underwent menopause at the conclusion of their reproductive careers and concealed their ovulation at the onset, how were men to know when a woman was at her peak reproductive state? Maybe by her breasts.

Crucially important for what we dub the Goldilocks hypothesis is the fact that such a signal would be difficult to fake: undeveloped breasts clearly indicate sexual immaturity, whereas sagging breasts indicate age. If so, then men should prefer breasts that are relatively plump—hence, providing comparatively honest information as to sexual maturity—and, as with the symmetry situation, women would to some extent be constrained to go along, especially insofar as the most desirable men (those likely to contribute positively to a woman's reproductive success) exhibited such a preference.

After acknowledging the possible role of prominent breasts as an indicator of fat reserves, Marlowe notes that "if this were the whole story, men should find breasts no more erotic than fat anywhere else on a woman's body." In addition, he asks, "How did breasts become attractive at all if hominid females originally had large breasts only when pregnant or lactating, and therefore not ovulating?"[35] In fact, as we have already pointed out, large breasts may well have become something for men to avoid if they indicated that a woman's reproductive services were already taken. But perhaps necessity was made into a virtue—or at least an opportunity for men—if large-breasted women were not so closely guarded by their existing consorts and if their bustlines, moreover, served as "continuous advertising" signals indicating that they were available for clandestine, extramarital, or nonmarital trysts. (Note: this explanation points in almost precisely the opposite direction as the provisioning hypothesis, which posits that full breasts stimulate male attention because they signal nursing and thus a positive payoff to male solicitude.)

As it is, many women comment ruefully on the unwinnable battle between their breasts and gravity played out with the passage of time. According to the Goldilocks hypothesis, it may be precisely for this reason that women have evolved protuberant nonlactating breasts: if men prefer such breasts because they cannot be faked (at least, in the long human evolutionary past that preceded the invention of push-up bras, other contraptions, and plastic surgery) and are therefore accurate indicators of age. Males should predictably be turned on by breasts—and by their owners—that are not too young, not too old, but just right, and that are prominent but also perky.

No surprise here that with the advent of cosmetic surgery, many women avail themselves of the opportunity to "correct" various "figure faults," notably breasts that are too small, too large, and so forth, or that in the process the most popular forms of physical fakery involve interventions that—like lip augmentation or eye wrinkle removal—mimic youthfulness and bodily health.

It is intriguing that a review of an enormous database consisting of more than 345,000 British and American works of fiction from the sixteenth, seventeenth, and eighteenth centuries found that romantic references to female breasts were strongly biased toward noting their "roundness," suggesting that a shape correlated with youth also correlated with erotic appeal.[36] Such a preference would be especially predicted in a bipedal species; among quadrupeds, breasts necessarily hang down and therefore do not provide information as to age. Beauty, it appears, isn't simply in the eye of the beholder, but in the health and reproductive potential of the person being beheld. In short, people are predisposed to see someone as beautiful insofar as she meets certain evolutionarily based criteria. (Don't forget those supernormal releasers.)

As with the question of symmetry assessment, however, why should women have gone along with this process? It isn't necessarily in their interest to evolve traits simply to make it easier for men to make reproductive decisions. The answer may parallel the symmetry dilemma: by evolving discernible breasts when young, women might have attracted males in part by offering a guarantee as to their own current age, but at the cost of giving away their age later, as the years accumulate. Such a trade-off might well also have been selected for because of the payoff to breeding well and early. As a general rule, any trait that confers a benefit when young is likely to be under selection pressure that is much stronger than the pressure against a trait that proves a liability after the reproductive years. This rule seems to explain why genetically mediated diseases such as certain cancers or, more dramatically, Alzheimer's that manifest primarily in the elderly persist when otherwise they ought to be selected against.

When it comes to breasts, the larger they are, the more rapidly they will sag over time. Hence, it is easier to judge the age of a large-breasted woman than the age of one whose bosom is less "developed." This distinction in itself may have induced men to prefer larger breasts because they provide more reliable information. Firm, protruding breasts would have been a way that women could advertise their youth, thereby attracting a larger number of admiring males from which they could then choose. Older females would of course then have been better off having smaller breasts because this size would have helped them obscure their age. But there is no getting around the fact that connective tissue stretches and weakens with the years, and it appears to be anatomically impossible to go from large firm breasts when young to smaller, equally firm ones in old age. In addition, selection wouldn't work against a trait that conveys a sufficient benefit early in one's reproductive career even if it becomes costly later, when an individual wouldn't be able to breed in any event.

The evolutionary pressures on breast development have doubtless focused not on elderly women, but on young ones, signaling that a girl has become a woman. After all, adolescent girls develop breasts at about the same time that they begin to ovulate, a process that, as we have seen, is carefully concealed among human beings. Breast development is exactly the opposite: unconcealed. It seems a bit strange that men should be so obtuse as to need ripening breasts to read reproductive competence. Among other species, males have little or no difficulty telling who is and who is not sexually mature. But then again, our sense of smell is retarded as far as mammals go, and given that ovulation itself is notoriously concealed in our species (chapter 3), perhaps breast development has evolved to fill the gap and provide information—"not too young, not too old, but just right"—not otherwise available.

We would be remiss at this point if we didn't at least mention the unlikely possibility that breasts might be signals to other women rather than to men. If so, what might they be saying? Just as menstruation might be directed at other women (a notion we raised and discounted earlier), it is at least imaginable that the case is the same with breasts. Perhaps breast development enables young women to announce that they are no longer immature and deserve to be taken seriously. As with menstruation, however, it seems at least as likely that women (especially when young and relatively powerless) would be selected to obscure their situation rather than call attention to it.

Doughty Daughters?

Finally, for yet another take on sexual selection and the evolution of female breasts, let us return to the peacock's tail. We already noted that breasts, like peacock tails, are costly. Although breasts do double duty, conveying practical benefit as well as a likely signaling function, biologists are pretty much agreed that the peacock's tail serves for signaling only. But what does it signal? And insofar as breasts, too, are signals, can a peacock's fancy feathers teach us anything about the evolution of breasts?

For one thing, a bizarrely elaborated avian tail says something about its manufacturer: that it is a male and, moreover, sexually mature. Ditto for human breasts: they say that their owner is female and, moreover, sexually mature (also, perhaps, of discernible age). The peacock's tail conveys more, however. It says that the peacock who produced it is physically fit, at least enough to have cobbled together all those gaudy feathers. Similarly, a healthy pair of breasts makes at least two promises: that whoever generated them has stored enough fat to fill them up and—less honestly—that their swollen appearance bespeaks a guarantee of nourishment for future offspring. These and other possible messages are concerned in various ways with a promised payoff to be received by the next generation and thus, by extension, one ultimately available to individuals making choices in real time. But a peacock's tail—and, we strongly suspect, a woman's breasts—also offers some purer, more rarified insights into sexual selection. For starters, think about a peacock's sons—an examination that has parallel implications, we argue, for a woman's daughters.

In addition to the "real" evolutionary benefits available to a choosy peahen from a peacock's showy tail, there is a purely sexually selected payoff that is no less real—because it, too, operates via reproductive success—but that works entirely in the realm of mate attraction rather than offspring nutrition. Because peacocks don't provision their offspring, the only thing they provide is their sperm, which is why these animals show such exaggerated sexually selected traits undiluted by other practical considerations. Peahens who chose cocks with classy tails increase the chances that their sons will have tails that are similarly alluring to the next generation of peahens. As a result, although such peahens may not have more offspring, they can count on more *grandchildren*—the offspring of their highly attractive sons. This notion, originally proposed in 1979, was labeled the sexy son hypothesis.[37] Since then, it has been abundantly confirmed and, despite occasional controversy, well established.[38]

This hypothesis leads us to suggest the following: What about applying the sexy son hypothesis to human beings? Substitute men choosing women for peahens choosing peacocks, and focus on prominent nonlactating breasts rather than on fancy tails. Instead of sexy sons, think doughty daughters.

The doughty daughter hypothesis argues that men may have begun choosing bosomy women because of the "honest" payoff implied by abundant breast tissue. Once such a preference became established and (regardless of the cause) took on a life of its own, men who fell for women with developed bosoms were likely to produce daughters who were similarly endowed. As a result, these men would have been evolutionarily more successful because their daughters (like the peacock's sexy sons) grew up to be sexually enticing and thus reproductively successful young women who produced grandchildren.

Evolutionary geneticist R. A. Fisher, whom we encountered briefly when considering RRV, also made an important contribution to the theory of sexual selection. He pointed out that preference for a sexually selected trait, once established, might have resulted in a "runaway process," generating traits that are excessive by standards of immediate practicality. If sexy peacocks obtain more matings simply because they are sexy, they would sire sons who in turn would also be sexy and who would be differentially more successful with pea-hens than their less gaudy competitors would be. Runaway selection of this sort wouldn't run away altogether because eventually it would bump against limits imposed by ecological/survival selection (and thus by natural selection's bottom line). At some point, the reproductive payoff of a sexually attractive tail will be balanced by its attendant disadvantages; the resulting peacock's tail that biologists have puzzled over—remarkably elaborate by purely practical considerations, but offering an optimal mix of ecological and sexual costs and benefits—is the outcome. By the same token, we propose that the female breast we are puzzling over in this chapter—excessive and impractical, yet offering a balance of metabolic and sexual costs and benefits—may be under-stood in the same way.

The connection is not exactly a slam-dunk. It needs to be shown, for ex-ample, that bosomy women have bosomy daughters—a likely correlation, but not to our knowledge empirically confirmed as yet. Moreover, such an associa-tion must be genetically influenced and not simply the result of a family tradi-tion of breast-friendly nutrition (if such traditions even exist). In addition, one reason the sexy son hypothesis works so well for peacocks and other animals is that among most birds and mammals, males have a higher variance in re-productive success than do females. In other words, there is more variability in male than in female breeding prowess: some males have more than their share of offspring, whereas others have less. By contrast, there is less difference between the most reproductively successful females and the least. A bull elk harem master, for example, whose retinue consists of ten females is likely to have ten offspring; each cow elk is likely to have one offspring, regardless of how many bulls she copulates with.

Males therefore tend to be the sexually selected sex because the prospect of high success (combined with the consequences of failure) generally acts more strongly on males than on females. Given the down side of generating a gaudy tail, it doesn't pay peahens to try to generate one because they will probably get the same number of eggs fertilized in any event. Seducing additional males isn't likely to enhance their breeding success, so it's better to spend their time and energy in more immediately useful ways. But because a highly desirable peacock can sire many offspring, whereas a drab one might remain forever celibate, the pressure is on males to be elaborately ornamented.

Why might human beings be different? In some ways, they aren't. Like the males of so many other species, men typically have a higher variance in reproductive success than do women; biologically speaking, *Homo sapiens* have evolved as mildly polygynous (harem-forming) creatures, which is why men are statistically larger than women, more inclined to aggression, and somewhat delayed in reaching sexual maturity. For the males in a polygynous species, it is distinctly disadvantageous to enter the competitive fray if you are too small, too meek, or too young. In contrast, because harem females are much less subject to these competitive pressures than is a wannabe harem male, there is no reason for such females to delay breeding and every reason to breed as early as possible. This difference is dramatically evident in harem-forming animals such as elephant seals and elk, as well as among humans in the early high school grades, when pubescent girls often tower over their later-maturing, opposite-sex classmates.

At the same time, people are unusual mammals in that they engage in biparental care (recall the possible benefits of concealed ovulation suggested by the keep him close hypothesis). Not surprisingly, we are also unusual in providing lots of paternal care; no other species of mammal has males that contribute so much to rearing successful offspring. As a result, it behooves women to be attractive to men, and not vice versa, as in most traditional evolutionary models of mate selection. The greater the significance of male investment for the success of offspring, the more important it is for females to signal their quality to any prospective mates. Growing full breasts (regardless of what may have initiated this trend) would therefore not only be appropriate for human beings, but more so for them than for any other mammals. Because men have something of value that they bring to the mating game, they, too, get to be choosy (traditional evolutionary models of mate selection emphasize female choice, even as they focus on male-male competition). Breasts, at least according to the doughty daughter hypothesis, are among the characters chosen.

When it comes to sexy sons and the like, Israeli zoologist Amotz Zahavi and his colleagues have come up with an intellectually challenging and counterintuitive suggestion known as the "handicap principle."[39] They propose that many traits used in sexual selection are actually liabilities when it comes to an animal's actually going about its life activities; moreover, it is precisely because these traits are disadvantageous that their possessors become sexually desirable! Presumably, any peacock can survive, but it takes a really exceptional one to do so while lugging around such a costly and troublesome tail. As a result, a hen who chooses an elaborately endowed cock is guaranteed to be getting some high-quality genes in the bargain, genes that will benefit her offspring; she will not just produce sexy sons, but also assure that her sons will inherit their father's overall vitality.

According to the handicap principle, we might therefore expect living things paradoxically to compete among themselves and to do foolish, wasteful, even dangerous things just to prove that their quality is such that they can get away with doing so. Nor are human beings exempt: along the Pacific Northwest coast of North America, chiefs of some Native American tribes used to engage in "potlatch ceremonies" in which they gave away and in some cases even conspicuously burned valuable food, baskets, and blankets to demonstrate that their wealth was so great that they could afford to waste it. Of course, in the process, they humiliated any rivals unable to match them in their extravagance.

Extending the handicap principle to human anatomy might go as follows. Breasts, beyond whatever is strictly necessary for lactation, may be a liability to their possessors, but what if this liability is precisely the point? A woman who shows herself capable of surviving and prospering despite her mammary handicap—who is, in a sense, so genetically "wealthy" that she can grow wasteful breasts for no survival benefit at all—must be a quality individual indeed. If, as a result, she is chosen by high-quality males, then her daughters will likely be similarly endowed—not just with breasts, but also with the ability to flourish in spite of them.

The sharp-eyed reader might notice a logical difficulty here. Insofar as a handicap (whether a peacock's tail or a human's breast) is a genuine liability, then how can it benefit an individual to inherit it, considering that along with the presumably superior genes that enable its possessor to transcend the liability, there also comes the handicap itself? A thorny problem, this, and resolving it has occupied dozens of mathematically inclined biologists. Current wisdom is that the possible evolutionary benefits of a handicap, however oxymoronic that sounds, may nonetheless be real.[40]

There is another route to success via the handicap principle. Let's grant that the same-sex offspring of the sexually selected sex (sons of peacocks, daughters of women) are stuck with a handicap along with whatever overall genetic quality it might signal. The real payoff may accrue less to them than to any *opposite*-sex offspring. Thus, when a peahen chooses a fancily feathered peacock, her (and his) daughters get the ability to flourish without the handicap of the tail itself. By the same token, when ancestral men chose elaborately ornamented women (i.e., those with conspicuous nonlactating breast tissue), their sons would have gotten the ability to flourish without the handicap of having to cart around those awkward mammaries.

Taking all of the foregoing into account, how did women get their breasts? Good question.

Fine-Feathered Females

Another related question also presents itself here. Breasts aside, why are women so fancy and men so plain?

In most species, it's the other way round: males are overdressed, and females dowdy. At one point in the musical *Hair*, a middle-aged woman sings out in support of men's long hair and other "flamboyant affectations" associated with the late 1960s hippie culture:

I would just like to say that it is my conviction
That longer hair and other flamboyant affectations
Of appearance are nothing more
Than the male's emergence from his drab camouflage
Into the gaudy plumage
Which is the birthright of his sex.
There is a peculiar notion that elegant plumage
And fine feathers are not proper for the male
When actually—
That is the way things are
In most species.[41]

Think of birds in general or of a particular type of bird—mallards, for example, in which drakes are ornamented with glossy green heads, whereas demoiselles are downright drab. Upon seeing a female duck, sparrow, hummingbird, or warbler among other similar birds, a nonornithologist may not be able to tell the species; males, by contrast, are generally distinctive, brightly colored and readily identifiable. The same applies to most coral reef fish, mandrill monkeys, and, indeed, most animals.

Male-male competition is one of the two major components of sexual selection. In our discussion of breast evolution, we have been concerned with the other primary driver of sexual selection: mate choice, especially males' presumed role in choosing females. Male gaudiness, in contrast, appears to pivot on the more frequent subset of mate choice—namely, females choosing males.

Just as the fundamental distinction between eggs and sperm drives males to compete with each other, it has also endowed females with something to be competed over—their large, nutrient-packed eggs and, in the case of mammals in particular, the prospect of subsequent female investment in the form of pregnancy and lactation. As a result, current evolutionary theory suggests that females are often in a position to choose among eager male suitors, the results being many of the extravagant male traits so apparent among so many animal species and females' selection of the healthiest and most desirable—that is, the "sexiest"—among an array of eager-to-oblige sperm donors. That is indeed "the way things are, in most species."

When it comes to the basic anatomy of naked human beings, however, neither sex is gaudier, more elaborately ornamented, or grotesque. *Homo sapiens* are sexually dimorphic, which is to say there is typically no problem telling male from female. At the same time, men and women are equally fancy or—depending on one's viewpoint—equally plain. Yet in most human societies it is the women, not the men, who are especially concerned with their looks. Even though American and European men occasionally spend time, energy, and money on their dress and appearance, women outdo them by an order of magnitude. In most human societies, women—not men—typically sport the equivalent of a peacock's tail, but one constructed culturally rather than anatomically.[42]

This pattern, of course, is far from universal. In some New Guinea societies, for example, men adorn themselves with pig tusks and bird of paradise plumes, and among the aboriginal people of Africa and Australia, men, not women, are elaborately painted for an array of ceremonies. Of all the "womanly mysteries" considered in this book, female adornment is the most culture bound and, in a sense, the least biological. To some extent, however, *any* human trait—insofar as it is found among any members of the species *Homo sapiens*—is necessarily "biological." And there is no question that when it comes to Western societies, the pattern "women fancy, men plain" not only goes counter to most animal species, but also qualifies as what scientists call a "robust phenomenon."

Just consider the expenditure on women's cosmetics (nail polish, lipstick, eyeliner, hair care products of all description) and clothing (brightly colored dresses, shoes, handbags, jewelry) versus the expenditure on men's cosmetics (aftershave, cologne) and clothes (shirts, pants, suits). Even feminist scholars such as Susan Brownmiller, Carol Tavris, and Naomi Wolf,[43] who typically do not endorse a biological interpretation of their findings, nonetheless agree that there is generally a greater emphasis on female beauty than on its male counterpart and, moreover, that women are typically far more likely to augment their physical appearance with "accessories."

This female-male difference with regard to beauty and body adornment is clearly influenced by cultural tradition, just as it is achieved via nonbiological intervention. Nonetheless, fancy females are for the most part a cross-cultural universal, with women depicted in consistently decorative roles.[44] An evolutionary perspective suggests that the cosmetics and fashion industry did not *create* a demand for female ornamentation so much as *respond* to it.

What might be the underlying cross-cultural biology responsible for this demand?

One hypothesis is that women are attempting to generate male demand . . . for themselves. After all, human beings are unusual among animals in the amount of "parental investment" provided by men, which, in turn, has placed a premium on women's making themselves—literally—attractive to potentially desirable mates, rather than vice versa, as in most animals. Even though monogamy is legally and ethically mandated (at least in Western countries), the fact remains that men are biologically primed to have a wandering eye to match their wandering gametes. This predisposition might help motivate women to attract any such wandering eyes; moreover, it is well known to evolutionary biologists that females of most species are inclined to prefer those males that have the most "goods"—good genes, good behavior, and good resources. Human beings may be unusual among living things in being able to augment their mate-selection strategies by conscious choice, which may include the choice to make themselves attractive and desirable to a potentially "good catch" and to do so, if possible, via artificial adornment.

This explanation, in turn, leads to a pair of "whys." First, why should women seek to emphasize certain traits rather than others? Probably because human beings are extraordinary in experiencing dramatic age-related changes in fertility (including menopause). It is noteworthy that in much of the world people are drawn to cultural practices and technology that includes breast enhancement, lipstick, hair and skin products, clothing that "flatters the figure" (or obscures various "faults"), and that all these interventions tend to exaggerate traits characteristic of healthy youth, along with its promise—albeit unconsciously communicated—of reproductive potential.

Second, why do women—and not females of other species—engage so enthusiastically in this sort of thing? One answer, of course, is that halibuts, hippos, and howler monkeys simply don't have the technology to apply mascara or to wear high-heeled shoes. But that probably misses the point, which is that human beings are different in another, crucial way: they are to some extent sex-role reversed compared to other mammals.[45] Not completely reversed, mind you: women—not men—get pregnant, give birth, lactate, are guaranteed to be genetically related to their offspring, and typically do more mothering than fathers do fathering. Nonetheless, among human beings, males make a substantial contribution (and not just as sperm donors) to their offspring's success, which means would-be mothers should accordingly be predisposed to choose those males most able and/or inclined to make such a contribution.

It is notable that among those few animal species in which sex roles are dramatically reversed—in which males make the greater biologically mandated parental investment—males are also the choosy sex, and females are the chosen. For example, among the dubiously named "Mormon crickets,"[46] mating males provide a huge glob of nutrient-rich material known as a spermatophylax, which the female consumes. Accordingly, male Mormon crickets are a good catch for females, who aggressively court them. The males, in turn, drive a hard bargain, holding out for females who promise a good supply of eggs, which they evidently determine by assessing body weight: as part of their courtship and mating ritual, female Mormon crickets must climb on the male crickets' back, who reject any females that are too light.[47]

Thus, even in this sex-role-reversed species, the basic pattern remains intact: the greater the reproductive contribution by one sex, the greater the choosiness of that sex and the greater the efforts by the other to show itself worthy of being chosen. Human beings are unusual in that both sexes make a substantial contribution—human fathers are at least capable of doing so—which, in turn, might help explain why women aren't merely choosy, but also seek to be chosen, and why, as a result, they often resort to adornment, creams, emollients, hair-dos and don'ts, depilatories, eye shadow, lipstick, rouge, powder, fingernail polish, and eye-catching dresses, bracelets, necklaces, and earrings. And why men notice.

As a general rule, the more monogamous the species and the more evenly balanced the parenting roles, the more similar are male and female. In polygynous (harem-forming) species, males are not only larger and more aggressive, but also proportionately more ornamented and adorned. By the same token, among "reverse harem," polyandrous species in which one female mates with multiple males, it is the females who are larger, more aggressive, and more ornamented and adorned. The harem-keeping sex also typically becomes sexually mature at a later age, when such individuals are more likely to succeed in competition against others of their sex. Women are smaller than men and less aggressive; they also mature earlier. As already discussed, these traits, plus the widespread prevalence of multiple wives rather than multiple husbands, makes a powerful case that human beings are biologically inclined toward polygyny.

But just as biologists were surprised when DNA fingerprinting revealed that females of most species seek (and obtain) "extrapair copulations,"[48] maybe biologists will also discover that women's public penchant for adornment corresponds to a private fondness for polyandry (multiple husbands). In short, maybe part of the reason women like to dress up—when given the chance and not forced by men into purda or a burqa—is that they like to attract more than one man. What's sauce for the gander may also work for the goose.

This proposal leads to another perspective on concealed ovulation, discussed in chapter 3: What if, rather than women's having been selected to conceal their ovulation, chimpanzees and the like have been selected to signal theirs conspicuously? Why should evolution have favored a female chimp's announcing her sexual availability? An obvious possibility is that the signal is meant to attract the dominant male's attention. This tactic would seem to be especially adaptive among species in which male genetic quality and/or inclination to be a good parent varies substantially among individuals. If there is a big difference among males, females should be selected to increase the chances of "getting" the better ones.

So what about the hypothesis that women, biologically more camouflaged, elect to use culturally created ornaments to achieve the same effect as chimpanzees do with their flagrant and presumably fragrant anatomy? This would make particular sense if women's penchant for polyandry is recently developed, so that natural selection hasn't had time to evolve overt physical traits such as the chimpanzee's gaudy genitals or it hasn't done so because of the various payoffs that come with being biologically more discrete.

Signals that enhance women's attractiveness might well include lipstick and rouge, which mimic the flush of sexual excitation; makeup that makes the eyes appear larger (belladonna has long been used in some cultures to dilate the pupils, thereby giving the impression of greater interpersonal interest); skin and hair products that enhance indications of youth and health; and corsets, stays, girdles, and push-up bras that modify the WHR. In addition, wearing expensive ornaments such as rings, necklaces, bracelets, and designer clothing may advertise and accentuate the wearer's financial value and thus her breeding prospects. We suggested earlier that several aspects of female anatomy and physiology such as concealed ovulation and prominent breasts may constitute evolution-based deception. It would seem that culturally mediated ornamentation offers even greater opportunity for various kinds of dishonest signaling. ("Dishonesty" in this case means simply the conveyance of information that is inconsistent with one's biological reality.)

Writing on "sexual selection and human ornamentation," biologist Bobbi Low points out something obvious, but nonetheless important: the distinction between signals and symbols. Whereas "signals such as dilated pupils or heightened cheek color are unlikely in any society or at any time to be interpreted as signs of sexual indifference, or indeed as anything but signs of sexual interest or excitation, . . . more abstract symbols, such as hair shades, for example, do not seem to relate directly to sexual fitness, suitability, or availability, and can have opposite meanings in the same circumstance, either in different societies or in the same society at different times. We may regard blondes as pure and virginal, or as wicked and likely to 'have more fun.'"[49]

Just as biologically generated traits have fitness costs—it is metabolically expensive and maybe more dangerous for a peacock to grow his fancy tail—it is often financially costly for people to ornament, adorn, and augment their bodies. But such costs are evidently seen as worthwhile: clothing, cosmetics, and other personal adornment compose a large proportion of the budget of many people who might seem to have "better things to do" with their limited resources. For a species that is deeply, biologically concerned with sexual signaling, there can be evolutionary logic behind such expenditures.

Out of 138 societies that Low investigated for aspects of male and female ornamentation, monogamous ones—as expected—exhibited significantly less ornamentation than did polygynous ones. In addition, 102 of the 138 used such devices to distinguish the marital or maturational status of *women*; by contrast, only 87 societies distinguished men's status by culturally produced ornaments, and nearly one-half of those did *not* indicate marital or maturational status. The researcher's conclusion: "There is a trend for female ornamentation to distinguish sexual availability (pubertal or marital status), while male ornamentation tends to distinguish rank and frequently puberty, but seldom marital status."[50] Think of the man who removes his wedding ring when on a business trip, but who brings along his platinum credit card, and of the woman seeking a husband who wouldn't be caught dead without her makeup.

Culturally generated styles and fashions can also accentuate conspicuousness, thereby offering a different take on attractiveness. They may, for example, offer the option of going counter to prevailing trends: when "everyone" is wearing long skirts, someone who desires extra attention—and who has a figure that permits it—might switch to shorter ones, and vice versa. Thus, although being stylish is a way of indicating that one is socially aware, capable of fitting in, economically competent, and otherwise "with it," being somewhat avant-garde says something different and a bit more daring. As a result, although most biological signals are likely to change slowly, determined by differential reproduction and the comparatively glacial pace of biological evolution, culturally mediated symbols will likely keep fluctuating in "real time" so long as people have something to communicate nonverbally. And so long as others can make money from their efforts.

When it comes to communicating nonverbally, there is also the matter of orgasm. As we discuss in the next chapter, orgasm may or may not have anything to do with communication, adaptive significance, maximizing fitness, or the other mainstays of evolutionary hypothesis making.

I t used to be conventional wisdom among biologists that human beings are unique in experiencing female orgasm, but no longer. Nonetheless, female orgasm remains both a marvelous phenomenon and a contentious, unsolved mystery among evolutionary biologists. Given the longstanding and widespread sexual repression of women in both Western and Eastern societies, it is not surprising that only recently has anorgasmia (failure to experience orgasm) been identified and treated. Nonetheless, the real biological mystery isn't why some women don't climax, but why some *do*.

5

The Enigmatic Orgasm

Aside from the scientific controversy it has generated, the likelihood is that female orgasm—even more than its fellow evolutionary enigmas menstruation, breasts, and concealed ovulation—has contributed mightily to making women sexually puzzling, often to themselves and certainly to men. Male orgasm pretty much speaks for itself. By contrast, female orgasm, because it is paradoxically both powerful and elusive, has long confounded efforts to see women's sexuality for what it is, causing men (and sometimes women) either to overestimate female eroticism or to devalue it. Because women—far more than men—are capable of multiple orgasms and also prone to anorgasmia, they have been seen as occupying one extreme or another: either being insatiable or lacking sexual desire altogether. Talmudic scholars once entertained such an overblown estimate of women's sexuality—and society's responsibility to repress it—that widows were forbidden to keep male dogs as pets! Yet, as anthropologist Donald Symons puts it, "The sexually insatiable woman is to be found primarily, if not exclusively, in the ideology of feminism, the hopes of boys, and the fears of men."[1]

Equally exaggerated and inaccurate, an influential nineteenth-century Victorian tract by one Dr. William Acton, titled *The Functions and Disorders of the Reproductive Organs, in Childhood, Youth, Adult Age, and Advanced Life, Considered in the Physiological, Social, and Moral Relations,* announced that "the majority of women (happily for society) are not very much troubled with sexual feelings of any kind. What men are habitually, women are only exceptionally."[2]

It has been said that the British and Americans are two people divided by a common language. Women and men are two sexes united *and* divided by their sexuality. And here orgasm looms large.

The "reason" for male orgasm seems obvious: no orgasm, then no ejaculation, no sperm transfer, no fertilization, no evolutionary success. By contrast, it isn't at all clear that female orgasm contributes to much of anything other than simply feeling good, which is another matter. "Pleasure," in evolution's purview, is something that reinforces behavior that contributes to fitness, just as pain or discomfort signals something to avoid. Insofar as it pleasurable to eat when hungry, sleep when tired, scratch when itchy, it is because eating, sleeping, and scratching enhance the fitness of individuals doing so, with "pleasure" a proximate mechanism whereby natural selection induces people to do what is biologically in their interest. So it avails nothing to attribute *any* pleasurable experience to pleasure itself, disconnected from an evolutionary payoff.

The problem is simply this: there is no evidence that orgasmic women make more babies or better ones. In fact, nonorgasmic women are quite capable of conceiving, as are nonorgasmic females of any species. Moreover, the success of artificial insemination, in human beings no less than in other species, makes it clear that sexual climax isn't necessary for fertilization. Why, then, does it exist at all?

Some Silly Suggestions

Before getting into the more plausible possibilities, let's dispense with some of the sillier ones. The redoubtable Desmond Morris—whose fertile imagination also brought forth the buttocks mimic hypothesis for breast evolution—came up with another whopper, which is that orgasm is nature's way of keeping a woman horizontal for a while after sex, thereby making fertilization more likely. This notion is regularly trotted out in scientific discussion of orgasm, mostly for historical reasons because before Morris, no one had even speculated as to its adaptive significance.

It has never been demonstrated, however, that remaining horizontal actually increases the likelihood that sperm will meet egg, and, as already noted, there is no evidence that orgasm predisposes toward fertilization. In addition, if staying on one's back or side enhanced fertilization, one can easily imagine any number of proximate mechanisms that would keep a woman lying down (reducing postcoital blood pressure, for example). Yet another problem is that if there is a fertilization payoff to remaining prone after copulation, animals that walk upright should also be subject to female orgasm for the same reason, yet kangaroos and wallabies pop right up and hop about immediately after copulating.

It might also be suggested—although we aren't aware of anyone who has done so—that anything that keeps a woman inactive after sex might render her less conspicuous to predators who may have been attracted to all that commotion in the bushes.

In 1829, Rev. Francis Henry Egerton, eighth and last earl of Bridgewater, specified in his will that eight thousand pounds be used to support publication, by the Royal Society of London, of his writings "on the power, wisdom and goodness of God as manifested in the Creation." The resulting *Bridgewater Treatises* include many purported examples of divine beneficence reflected in the natural world. For example, the goodness of God is demonstrated by the fact that each rattlesnake is outfitted with a rattle by which people are warned of imminent danger. At the same time, recourse to such "reasoning" must also deal with the other end of the rattlesnake! Similarly, the sex as silencer hypothesis would have to contend with the other end of a sexual encounter: granted that there may be a postorgasmic diminution of noise, but orgasm itself—female no less than male—is often pretty noisy. If our ancient ancestors were being prodded by natural selection to avoid detection (by other conspecifics, perhaps, no less than by predators) during or after sex, they would seem well advised to have kept quiet about it.

A Copulatory Carrot?

Our initial reaction to another superficially plausible hypothesis, the idea that maybe female orgasm has simply evolved as a way to get women to copulate, is to dispense with it. The notion is that intercourse takes time and trouble, is sometimes uncomfortable and even dangerous, so ancestral women needed to be encouraged via an orgasmal carrot. The main problem here is that lots of other animals—female as well as male—engage in sex with no indication of anything remotely approaching orgasmic bliss. Rather, they copulate with the same kind of bored resignation—and occasionally enthusiastic insistence—that they bring to other life activities such as eating, grooming, building a nest, patrolling their territories, and so forth. Orgasm clearly isn't necessary in order for other animals to copulate, although some sort of "positive reinforcement" seems in order, just as it makes sense for living things to find food satisfying.

Most animals, however, don't need profound waves of cataclysmic ecstasy in order to get them to eat. Why should sex be any different? Indeed, it isn't, which is why biologists were not especially perplexed that other animals copulated without appearing to need female orgasm as a motivator. But maybe we are moving too quickly in discarding the copulatory reward hypothesis. After all, just as pregnancy and childbirth are inconvenient, often painful, and even dangerous, so is sexual intercourse. And just as Nancy Burley's consciousness hypothesis suggested that our ancestors may have sought to avoid sex when they were most fertile,[3] thereby selecting for concealed ovulation insofar as those who did not know when they were ovulating were likely to have *more* offspring, maybe at least some of our ancestors—especially our ancestral mothers—sought to avoid sex altogether.

Let's face it: heterosexual intercourse is a hassle. It takes time and energy, distracts the participants from other more immediately profitable things they might otherwise do, carries risk of disease as well as physical injury from a possibly overzealous or otherwise uncooperative partner, and requires often difficult-to-achieve coordination with someone else who is likely to have a separate agenda. Given all these downsides, isn't it at least possible that women might need particular rewards in order to engage in sex? Carl Sagan used to say that extraordinary claims require extraordinary evidence. By the same token, perhaps the extraordinary costs of heterosexual intercourse required extraordinary immediate payoffs. To be sure, the ultimate, eventual biological payoffs—bearing children and thus achieving a degree of evolutionary fitness—are clear enough. But maybe they aren't cogent enough in "real time" to compensate for the liabilities of intercourse, especially for a species such as *Homo sapiens,* endowed (or burdened) with consciousness and thus liable to have second thoughts about something that happens "naturally" among other creatures.

Just as Nancy Burley's consciousness hypothesis relies on human self-awareness (and paradoxically would involve selection for ovulatory *un*awareness), the copulatory reward hypothesis suggests that ancestral women who weren't outfitted with an orgasmic reward may not have had sex—or simply may have "had" less—than did those more orgasmically capable. If so, then as with the consciousness hypothesis for concealed ovulation, selection would have favored the evolution of orgasm because those women who climaxed would have been more likely to engage in more intercourse and presumably to produce more children.

We are intrigued by this hypothesis, but somewhat skeptical nonetheless. Many women are anorgasmic but reproduce anyway. It would be interesting, however, to learn whether orgasmic women engage in *more* sex and—crucially— whether they produce *more* offspring. As for the fact that female orgasm is notoriously fickle, psychologists long ago established that "intermittent reinforcement"[4] (a reward that isn't wholly consistent) is often more effective in producing a given behavior than one that is wholly reliable. It also seems worth considering whether the same basic argument might apply to men because the immediate costs of sexual intercourse are substantial for them, too, and perhaps the intensity of male orgasm, no less than that of women, has evolved specifically as a compensating copulatory reward.

There also remains the problem of why so many other animal species copulate willingly, often eagerly, but without apparent orgasm, despite the fact that for most of them the costs of intercourse are likely as great as they were for our ancient human ancestors. Maybe it's once again a matter of consciousness in that other animals, lacking conscious awareness of the costs of copulation, don't need the powerful motivation of orgasm, whereas our forebears, being in a sense too smart for their own good, did.

In any event, it now seems that the females of at least a few other species do in fact experience orgasm, at least as indicated by certain physiological measures such as sudden elevation in respiration and heart rate, muscular contractions, convulsive vocalizations, and so forth.[5] Which leads to the inevitable question, what *is* female orgasm anyhow?

Commenting on pornography, Supreme Court Justice Potter Stewart made the immortal observation: "I may not be able to define it, but I know it when I see it." Female orgasm is about as difficult to define. Here is one effort that appeared in a recent major scientific review: "An orgasm in the human female is a variable, transient peak sensation of intense pleasure, creating an altered state of consciousness, usually with an initiation accompanied by involuntary, rhythmic contractions of the pelvic striated circumvaginal musculature, often with concomitant uterine and anal contractions, and myotonia that resolves the sexually induced vasocongestion and myotonia, generally with an induction of well-being and contentment."[6]

Let's simply note that most women know it when they experience it. All those contractions, all that myotonia (muscle relaxation after contraction), and all the resolved vasocongestion (variations in local blood supply) aren't important merely because they are the physiological correlates and possibly the causes of the subjective sensation, but also because they allow biologists to be pretty confident in asserting that a few other species experience something akin to the human version. But the fact that female orgasm occurs in some other species doesn't negate the fact that it is absent in most sexually reproducing animals. Yet reproduce they do.

Just because something works in a particular way or can be made to do so doesn't mean that it was intended for that purpose. Literary theorist Viktor Shklovsky, in a renowned essay titled *Knight's Move,* pointed out that "[i]f you take hold of a samovar by its stubby legs, you can use it to pound nails, but that is not its primary function." He went on to note that during the Russian civil war, "With my own hands I stoked stoves with pieces of a piano in Stanislavov and made bonfires out of rugs and fed the flames with vegetable oil while trapped in the mountains of Kurdistan. Right now I'm stoking a stove with books. . . . But it's wrong to view a samovar with an eye to making it pound nails more easily or to write books so that they will make a hotter fire."[7] Almost certainly, it is equally wrong to view the female orgasm as designed to induce sexual intercourse just because it may provide an occasional payoff for doing so.[8] The question of orgasm's "primary function" is still an open one.

Orgasm is a great reliever of tension. At a subjective level, that is how many people describe the experience: like floodgates opening, a spring uncoiling, an explosion of pleasurable relief. But this doesn't mean that tension reduction is the *reason* for orgasm. After all, sexual tension wouldn't accumulate and need releasing if it hadn't built up in the first place; hence, it is circular to maintain that orgasm exists so as to get rid of something (tension) that wouldn't otherwise be there—disconcertingly like banging one's head against a wall because it feels so good when you stop.

Finally, yet another "explanation" for female orgasm turns out upon inspection to be no explanation at all: it is a gift from God or from evolution or both. We cannot scientifically evaluate the former possibility, except to note that although women are doubtless deserving of such a gift, we are skeptical that positing a divine sexual Santa Claus provides any explanation at all. Think of Meg Ryan's famous simulated-orgasm-in-the-restaurant scene in the movie *When Harry Met Sally* (after which a middle-aged fellow female diner says to the waiter, "I'll have what she's having").

We'll look at fake orgasm later; for now, our point is simply that the real thing cannot simply be written off as good fortune or as a previously unrecognized culinary consequence of choosing the right menu item while having a meal with Billy Crystal. But what about the possibility that it evolved as a result of evolutionary beneficence, disconnected to any fitness payoff—in other words, pleasure as its own reward, for its own sake? If this turns out to be true, then it will be the first and probably the only case of its kind. Evolution is a rigorously natural phenomenon, tightly coupled to cause and effect. The causes are sometimes random and unpredictable and sometimes themselves a result of uniquely contingent historical events; most of the time they derive from certain traits that convey a reproductive edge, however slight, over their alternatives. But never, nowhere in the natural world, has any gift—however pleasurable—been bestowed out of mere cosmic generosity.

An Evolutionary By-Product?

Let's turn now to the most challenging scientific explanation for the existence of female orgasm—challenging, that is, to evolutionary thinkers because it essentially discounts the role of evolution itself, or, rather, the role of adaptation. An adaptation, for evolutionists, is a trait whose existence is due to the fact that it has been positively selected for, which is to say it seems likely to have evolved because it increases the fitness of those possessing it. Every characteristic of every living thing has to some extent passed through natural selection's biological sieve; if the characteristic is on balance fitness enhancing, it is likely to be preserved, whereas if its costs are sufficiently great, it will eventually be selected against and its alternative selected for (assuming that an alternative is available). But this does not mean that every characteristic of every living thing exists because it conveys a reproductive benefit, that all traits owe their existence to their success in somehow helping their associated genes project themselves into the future.

In his highly influential book *Adaptation and Natural Selection,* the great evolutionary theorist George C. Williams made a good point about flying fish.[9] Clearly, their ability to "fly" (actually, glide through the air) is an adaptation, resulting from selection for a fusiform body shape, wide aerodynamic fins, and the mechanical ability to swim very fast then launch into the air, thereby escaping predators such as tuna. Moreover, it is clearly biologically beneficial that after traversing long distances in the air, flying fish fall back into the water, considering that they can breathe only through gills and are adapted to live in water, not air. But—and here is the key point—the fact that flying fish eventually stop flying and splash back into the ocean is not in itself an adaptation. Evolution doesn't make these animals descend; gravity does. For these animals, "flying" is clearly an adaptation, but flying fish end up in a gill-friendly environment and thus capable of going on with their lives as a simple by-product of their physical existence, not as a consequence of any direct evolutionary process.

One more example. Noses are more than a little useful for holding up eyeglasses, but they didn't evolve with spectacles in mind. The beakish, protuberant quality of the human nose has presumably evolved for its own good reasons. Smell is clearly an important human sense, even though our olfactory prowess is minimal compared to that of most other mammals. Olfaction serves certain obvious biological functions, such as warning us of putrid or chemically polluted substances and attracting us to pleasant and healthful ones. It also serves less obvious but no less biological roles, such as sensitivity to the pheromones of which most people are unconscious and the generation of a potent link to memory, as exemplified by Proust's famous reminiscence over a tea cake. But the human nose's eyeglass-holding potential was not responsible for its evolution of, even though that, too, is something many noses do these days.

Is female orgasm like the waterborne descent of flying fish or the eyeglass-holding capabilities of the human nose?

The by-product hypothesis for female orgasm was first suggested by anthropologist Donald Symons in his 1979 book *The Evolution of Human Sexuality*, enthusiastically endorsed by Stephen Jay Gould and supported most recently by philosopher of science Elisabeth Lloyd in her book *The Case of the Female Orgasm*.[10] This theory holds that female orgasm is a "sexual vestige," something that persists in women simply because it is useful in men and can't easily be eliminated. Women have orgasms, in other words, because men do, and evolution hasn't gotten around to eliminating it, or maybe the cost of doing so is simply too great. (So, if you are a woman, enjoy orgasms while you can because they may be snatched away at any time—or more likely at any old eon.)

Evolutionists have no hesitation acknowledging that whatever trait they are studying might be nonadaptive (i.e., neutral) or even downright maladaptive, but in fact they don't usually take these options very seriously. When it comes to female orgasm, however, the nonadaptive possibility is especially pronounced, or, at least, it has received considerable positive attention, so it must be confronted.

Ornithologist David Mindell points out that evolutionists tend to treat nonadaptive explanations like most of us treat those annoying "terms of agreement" windows that pop up when we are installing new software on a computer. After scrolling through a mass of fine print (which none of us reads), we are expected to click on the ACCEPT button before proceeding. Then we often get another collection of boilerplate. Evolutionary biologists likewise routinely pay lip service to the mantra that their study subjects might very well exist *not* because they are adaptive in themselves and thus the result of natural selection favoring such traits, but because they are indirect consequences of evolution's favoring other things instead.

Here is Mindell's account of how this might work when it comes to nonadaptive interpretations:

"The appearance of various behaviors and mental functions as discrete, modular, or adaptive targets of selection may be artifacts of human perception, reflecting misunderstanding of developmental, genomic, physiological or environmental effects . . . " **ACCEPT**

"Current functions for traits can be very different from previous or original functions, weakening assumptions regarding their adaptive origins and maintenance. Traits and whole organisms may represent a hodgepodge of compromises with re-engineered components, resulting from inelegant, minimally harmful changes over time . . . " **ACCEPT**

"Chance events, including those of earth history, local environmental change and genetic drift, can and do play seminal roles in the differential success of individuals, populations and species exhibiting variations in behaviors of interest . . ." **ACCEPT**[11]

Finally an inner voice says, "OK, I've done the obligatory genuflections to the various nonadaptive possibilities. Now can I get on with the main event: looking into adaptive explanations?" Mindell's point is that evolution by natural selection is so powerful and its adaptive consequences so fascinating and wide ranging that it readily takes over most evolutionists' imaginations, leaving even the possibility of nonadaptive interpretations languishing in a limbo characterized by "accept in theory but do not embrace in practice."

The By-product Hypothesis for female orgasm is a nonadaptive interpretation if ever there was one. And although at first blush it might seem absurd—and demeaning of women as well—to claim that women's orgasms are merely an irrelevant reflection of the Real Thing, namely male orgasm, it isn't mere boilerplate fine print to be ignored before pushing the ACCEPT button. Advocates of this hypothesis point to a convincing mirror-image situation: the case of male nipples. Men derive no evident benefit from having nipples. The likelihood is very great that male nipples are a classic tag-along trait, a by-product of the fact that *female* nipples have been strongly selected for as milk delivery systems. Moreover, both male and female embryos experience a common developmental pathway, such that it would be a major fitness hurdle for selection to interfere and mess up a delicate and deeply embedded embryological sequence that works just fine, and carve out an unnecessary exception for men. After all, male nipples don't do any harm. Embryonic development is often a package deal, and male nipples are evidently an unavoidable part of the package.

If male nipples caused trouble—for example, if they were a frequent site of cancer or other disease, or even if they were metabolically expensive to produce—they would presumably have disappeared long ago. But they ain't broke (they're merely irrelevant), so evolution hasn't fixed them. Just as male nipples almost certainly owe their existence to the benefit of nipples when flourished by women, maybe female orgasm exists just because it is strongly selected in men. (And, like male nipples, it certainly doesn't seem to do any harm.)

The notion has plausibility. After all, just as male orgasm and ejaculation occur in response to stimulation of the penis, female orgasm is intimately linked to stimulation of the clitoris, and both the clitoris and the penis derive from the same undifferentiated embryonic tissue known as the genital ridge.[12] So perhaps it shouldn't be surprising that the two organs are richly endowed with nerves and that there are brain mechanisms that respond—orgasmically—to enough of the right input from those nerves. In one case (men's), the response is adaptive, whereas in the other (women's) it is essentially an evolutionary hitchhiker.

We frankly are a bit surprised that the By-product Hypothesis has received so much respectful attention from so many knowledgeable biologists. Perhaps they are seeking to demonstrate their scientific bona fides, to show that they have read the nonadaptive fine print, take it seriously, and are open-minded about its potential. Or perhaps they are too readily seduced by that male nipple analogy. "Analogies," as Sigmund Freud once wrote, "decide nothing, but they make one feel more at home."

They can also mislead. For one thing, male nipples don't *do* anything, whereas the clitoris does plenty. Men's nipples are small and inconspicuous, as befits a by-product upon which selection has been reduced. By contrast, female orgasm is complex, highly elaborated, and downright Technicolor compared to its comparatively feeble male counterpart. Rather than existing as an insignificant, pale imitation of where the genuine action is supposed to be (in men), female orgasm shows every sign of having been structured and fine-tuned by evolution. Anyone who has experienced or has been with someone who has experienced female orgasm in its multidimensional intensity and repeatability might well be reluctant to call male orgasm the Real McCoy, while relegating female orgasm to tag-along insignificance.

It is also possible that male nipples are adaptive after all, calling attention, perhaps, to male body symmetry or the lack thereof in precisely the same manner as has been proposed for their female counterparts—although without accompanying breast tissue, any such assessment would certainly be of diminished utility.

For another thing, women's orgasms are unlike men's nipples in that they aren't always there. There are no nippleless men, but lots of anorgasmic women. Nor are there any men who develop nipples some of the time, although there are lots of women—the great majority—who experience orgasm only sometimes. Male nipples, by all accounts an unavoidable consequence of human embryology, are persistent, albeit nonfunctional. Female orgasms are not consistent, but, as we shall see, this variability may paradoxically be part of their ultimate functionality.

Freud has been roundly and rightly criticized for declaring that clitoral orgasm is "immature" compared with its more "developed" vaginal counterpart. After all, the clitoris seems clearly "designed" by evolution to help generate orgasms. It has, by some estimates, more than eight thousand nerve endings, exceeding the number found anywhere else in the human body and approximately doubling that of the penis. Its only known function is heightened sexual sensation. If the clitoris (and thus the orgasmic consequence of stimulating it) exists merely as an unavoidable by-product of selection to produce the penis, why is it *more* neuronally endowed than the penis?

The human embryo's "default setting" is female, not male. In the absence of prodding by androgens, the fetus develops a vagina, clitoris, uterus, ovaries, fallopian tubes, and so forth. So is it legitimate to conclude, as do Symons, Gould, and Lloyd, that the penis and male orgasm are somehow the basic, infrastructural pair, the apple of evolution's eye? In fact, a case might be made that male orgasm isn't nearly as necessary as many people assume; there is no reason, for example, why ejaculation can't be as unexciting as, say, urination—something biologically necessary and therefore prompted by feelings of urgency but without any orgasmic Sturm und Drang.

It has also been claimed that women's orgasm can't possibly be adaptive because it isn't consistently evoked during heterosexual encounters. Indeed, according to a recent study, about 33 percent of women report never being orgasmic during sexual intercourse, compared to only about 20 percent never being orgasmic during masturbation; in other words, about 67 percent are orgasmic at least on occasion during intercourse, whereas 80 percent experience orgasm as a result of masturbation. Incidentally, this study focused on twins, and it also found that the wide variation in sexual function has a genetic basis and is not simply attributable to cultural factors: roughly 40 percent of the diversity in women's ability to reach orgasm is correlated with underlying genetic variation, about comparable to findings for other "genuine" biological traits such as migraine, hypertension, age at menarche, and menopause.[13]

Advocating for the By-product Hypothesis, Elisabeth Lloyd claims that orgasm's "phenotypic plasticity" is "evidence that selection has not acted on the trait at all."[14] (The term *phenotype* refers to any discernible characteristic of a living thing, such as its size, shape, color, or presence or absence; *plasticity*, of course, means the degree to which something is malleable, likely to appear in different forms.) But selection can favor phenotypic plasticity. Indeed, it often does, especially when operating on *Homo sapiens*. The fact that human beings speak thousands of different languages is testimony to how selection can lead to a general capability while at the same time varying the outcome depending on circumstance: if a child grows up with people speaking Chinese, she learns Chinese. More to the point, deaf infants do not develop normal speech because they haven't experienced the necessary auditory feedback. There is no question that natural selection has endowed our species with linguistic ability, but also that it has left the details open to the vagaries of experience.[15] The fact that there is wide "phenotypic plasticity" in the human capacity for language is perfectly compatible with the assertion that selection has acted on the ability to acquire and use language. By the same token, the fact that women differ in their experience of orgasm is not evidence that selection hasn't acted on the trait. In fact, as discussed more fully later, it may well have acted precisely to make orgasm variable not only from one person to the next, but within the same individual's lifetime experience.

Lloyd also points to the "puzzling data on the relative infrequency with which women experience orgasm with intercourse" (the rather inelegantly named "intercourse-orgasm discrepancy"), maintaining that "under the common assumption that the capacity for orgasm is designed as an adaptation to encourage and reward intercourse, this infrequency must be seen as a design flaw."[16] Any flaw is admittedly—by definition—a sign that something is wrong, but there is nothing about evolution by natural selection that precludes the existence of flaws! Evolutionists know that the biological world abounds in flaws: our narrow birth canal, the structure of our knees and lower back, the location of the prostate and of the exit site for the optic nerve. These flaws and many others are a result of past historical contingencies and the fact that evolution has never had the benefit of designing living things from scratch to be perfect. It is the advocates of "intelligent design" and of similar drivel who have difficulty contending with design flaws. Biologists understand that because there is no designer, flaws are to be expected. Organisms weren't designed *de novo*, but rather evolved to be what they are via small steps from what their ancestors were. If, however, a divine designer were responsible, then she must be either incompetent, indifferent, lazy, or, on occasion, downright malign.

Granted, orgasm is more readily induced by oral, digital, or mechanical stimulation of the clitoris than by a penis penetrating the vagina. This fact suggests, among other things, that evolution isn't always maximally efficient, not that orgasm isn't an adaptation. Evolution has to work with what is available, not with what might be optimal. For example, human beings' lower back is poorly designed, having evolved from a structure that among our ancient quadruped ancestors used to be parallel to the ground. By the same token, the prostate gland is a nightmare for aging men, once again a result of the fact that evolution didn't design the human body de novo with maximum engineering efficiency; rather, it produced a ramshackle affair, having been forced to build on preexisting structures and modifying them as available and as subject to the constraints of history and embryology.

Thus, it is indeed true that the clitoris is derived embryologically from the same tissue that in males gives rise to the penis. And the two organs send impulses to the brain in the same way: via the pudendal nerve. However, the following is also true and more than a little inconvenient for those arguing that female orgasm is merely a by-product of its male counterpart. There are three more pairs of nerves (pelvic, hypogastric, and vagus) that conduct sensation from the vagina, cervix, and uterus to the brain. Moreover, stimulation of these organs—conveyed to the brain via their respective nerves—can also induce female orgasm even without any mechanical stimulation of the clitoris.[17]

Women also describe that their subjective experience of such orgasms is different from that achieved via clitoral stimulation. It is also worth pointing to the obvious: there is a definite market for dildos, which wouldn't exist if vaginal stimulation in itself weren't perceived, at least sometimes, as highly pleasurable. All of this strongly suggests that female orgasms have their own anatomical and physiological underpinnings, almost certainly produced by evolution, which in turn makes it highly likely that they are of adaptive value in themselves and not simply irrelevant hitchhikers upon male orgasm.

At the same time, there is no doubt that when it comes to female orgasms, the clitoris is "where its at." If genital embryology were somewhat different, such that the clitoris were to find itself inside the vagina, then women would doubtless find themselves more inclined to climax as a result of vaginal penetration. But because of embryologic constraints, the genital ridge—which eventually differentiates into either clitoris or penis—doesn't end up there. Resourceful human beings—and several monkey species—have therefore discovered other, more efficient ways to stimulate it and achieve an outcome that is also available via other means. In short, the sad truth (at least, for heterosexuals) is that female orgasm is not well designed to be reliably elicited by penile-vaginal intercourse. Under various scenarios, it might be more adaptive yet and certainly more convenient if women were able to achieve orgasm by eating chocolate, checking their e-mail, thinking pure thoughts, and so forth, but insofar as orgasms are keyed to penises and clitorises, that's what natural selection has had to work with.

Yet the "intercourse-orgasm discrepancy" is somehow supposed to weigh heavily against the adaptive significance of female orgasm itself. By the same token, one might point out that male orgasms are more reliably achieved if a man stays home and masturbates than if he goes out on the town to seek a sex partner, but no one would seriously claim that this shows that male orgasm during intercourse isn't adaptive. A hungry person might be able to achieve a kind of satisfaction (albeit temporary) by filling her stomach with sand or chewing coca leaves, or she might force herself not to eat altogether, but this fact doesn't argue against the adaptive significance of consuming food.

Don't misunderstand us: we have nothing against masturbation or same-sex relationships or carnal satisfaction achieved via poetry, sunsets, cooking utensils, antique harpsichords, or even consenting animals. Quite the contrary. Let a thousand orgasms bloom! In a world both overcrowded and increasingly infected with dangerous sexually transmitted diseases, there is much to be said for masturbation in particular as the epitome of safe sex. The point is that just because something (e.g., female orgasm) can be achieved in diverse ways (e.g., masturbation, same-sex encounters) does not argue against its having evolved because it is particularly adaptive in a specific, different context (e.g., hetero-

EVOLUTIONARY HYPOTHESES EXPLAINING WOMEN'S ORGASMS

Facilitate fertilization 1: keeps woman lying down
Predator avoidance: postcopulatory quiescence
Copulatory reward: positive reinforcement
By-product: a tag-along consequence of its male counterpart
Infanticide insurance redux: a spur to multiple partners
Monogamy: pair–bond enhancement
Facilitate fertilization 2: uterine upsuck
Facilitate fertilization 3: reduction in flowback
Facilitate fertilization 4: biochemical benefits
Facilitate fertilization 5: sire choice
Mate evaluation 1: social and/or genetic quality
Mate evaluation 2: parenting potential

sexual intercourse). After all, that's the context in which all sexual species, including *Homo sapiens,* evolved.

Once orgasm became available to women, for whatever reason, it wasn't "nonevoluntary" for it to be achievable via a shortcut or two. Consider, for example, that human beings have evolved a fondness for sugars because simple carbohydrates characterize ripe fruit, a good food source for our tree-dwelling primate ancestors. Thanks to the confectionery industry, people can now pleasure their sweet teeth with nonadaptive candies, chocolates, and soft drinks, but this doesn't mean that the human penchant for sweetness doesn't show the hallmarks of natural selection. Masturbation or oral sex may be similarly sweet or even downright scrumptious without arguing against an adaptive value to intercourse-based pleasure.

This juncture may be where Freud went astray. His fixation on the vaginal orgasm as more "mature" than its clitoral counterpart was consistent with a simplistic if unstated evolutionary assumption: namely, that orgasm should somehow be associated with sexual intercourse and thus with reproduction. In this narrow perspective, vaginal orgasm makes sense; clitoral orgasm doesn't. Now, in a post–Masters and Johnson world, we know that clitoral orgasms are, if anything, more common, more readily evoked, lengthier, and more intense. Penis and vagina are a nice lock-and-key pair, satisfying both evolutionarily and personally. The fact that for women fingers or tongue upon clitoris are typically even more satisfying doesn't undermine the adaptive significance of either heterosexual intercourse or of the female sexual response, however evoked.

Many women don't achieve orgasm, but perhaps they would be able to in the right circumstances. Some presumably would not be able to at all. Again, this distinction wouldn't mean that orgasm isn't an adaptation. Most people can run and jump, but some can't, and of the latter, some probably never will. This variation doesn't mean that the capacity to run and jump isn't adaptive and that the suite of anatomy and behavior enabling most healthy people to do so hasn't been produced by natural selection. It turns out that women who experience sexual dysfunction are liable to have substantially reduced clitoral inervation via the pudendal nerve.[18] Accordingly, just as there are natural athletes as opposed to the rest of us—even though a degree of athleticism has almost certainly been selected for—there may well be natural sexual athletes, too, as opposed to those whose neuronal functioning is suboptimal. The variation exists not necessarily because evolution doesn't favor the higher performers, but rather because many biological traits are distributed across a continuum. Not everyone is an Olympic champion. Indeed, those women who achieve orgasm with every sexual encounter are probably less likely to be the focus of natural selection than are those who fall in the "sometimes" category, as we discuss later.

But, first, we must look at another argument promoted by advocates of the by-product hypothesis that, once again, has a patina of bio-logic. The claim is that traits that are important to fitness show very little variability, whereas orgasm is notably variable—among individuals, and also within each individual, from one experience to the next. Hence, orgasm can't be adaptive. Once more, there is a faint ring of truth to this conclusion. Having one head, for example, evidently is adaptive, and sure enough there is no variability in the human population when it comes to headedness. In contrast, eye color does not appear to be adaptive in itself, and there is plenty of variability on that score.

It is true that strong selection tends to winnow away genetic diversity, but in fact many things can contribute to variability in a trait, and underlying genetic variability is merely one of them. Intelligence (at least, in comparison with our nonhuman relatives), eyesight, disease resistance, and so forth have clearly been selected for. And yet these traits tend to vary substantially among individuals, in part because they are influenced by many different genes and in part because there are complex and varying interactions between each individual's genetic underpinnings and the environment that he or she experiences. For example, a study of orgasmic frequency among Swedish women from eighteen to seventy-four years old found generational differences to be more important than age as such: attitudes toward sex and toward one's own sexuality have a greater impact than does age or even sexual experience.[19]

Impetus for the by-product hypothesis has come from several sources: Stephen Jay Gould's eagerness to identify human traits that are not adaptive (in support of his peculiar insistence that evolution has had less to do with human nature than with the nature of other living things); the fact that female orgasm is—like the other traits covered in this book—a genuine evolutionary enigma in that its adaptive significance, if genuine, is certainly not obvious; as well as a scientifically legitimate desire to explore all possible explanations for any biological enigma of this sort, including the "null hypothesis" that it might not be a direct product of evolution after all.

Feminists in particular might be expected to object to the by-product hypothesis because it seems to marginalize female sexuality, demoting climax to a kind of frivolous add-on to its more genuine and functional male counterpart. Thus, it may well fall precisely into the widespread and long-standing sexist error Simone de Beauvoir identified in her classic manifesto *The Second Sex*: namely, it takes the male situation as the "Subject" while treating the female as tag-along "Other."[20] But if female orgasm were indeed such a by-product, then it would be incumbent upon biologists to accept it as such, regardless of whether the political implications might be regrettable. As it happens, the probability is otherwise.

At the same time, another subtext seems to be motivating at least some feminists' adherence to the by-product hypothesis. Insofar as female orgasm might exist as a by-product (even if a by-product of something in men), its significance can paradoxically be more readily separated from heterosexual intercourse and its elicitation thereby further legitimated in same-sex or autoerotic contexts.

Let's be clear again. We—Barash and Lipton—are not only sympathetic to, but downright enthusiastic about happy and gratifying sexuality, however achieved, whether through same-sex relationships—female as well as male—or masturbation. There is much to be said for obtaining and enjoying erotic satisfaction, including but not limited to orgasm, in as many ways as possible: from warm baths, stroking a cat, or being pleasured by a friend of same, opposite, or indeterminate gender. But whether we like it or not, the current burden of scientific evidence is (1) that female orgasm has indeed evolved and, moreover, (2) that it did so in the context of male-female sexual encounters, even though it is no longer limited to heterosexual intercourse or even most readily evoked in this venue.

How might that have happened?

Infanticide Insurance and Fertilization Redux

Anthropologist-primatologist Sarah Blaffer Hrdy suggests that female orgasm evolved as a spur to having sex with many different males. "Based on both clinical observations and interviews with women," writes Hrdy, "there is a disconcerting mismatch between a female capable of multiple sequential orgasms and a male partner typically capable of one climax per copulatory bout."[21] A potential consequence of this "mismatch" is that females would be inclined to seek multiple partners in order to achieve their orgasmic potential. As for why this potential exists at all, Hrdy suggests that it is ultimately driven by the fitness benefit of taking out an anti-infanticide insurance policy, as proposed earlier for the evolution of concealed ovulation. Thus, female orgasm and its requirement of sustained stimulation may have provided the proximate mechanism underpinning the ultimate payoff deriving from having sex with multiple partners. Here are Hrdy's own words: "It is possible that as in baboons and chimps the pleasurable sensations of sexual climax once functioned to condition females to seek sustained clitoral stimulation by mating with successive partners, one right after the other, and that orgasms have since become secondarily enlisted by humans to serve other ends (such as enhancing pair-bonds)."[22]

Picture humanity's mother studiously going from one sexual partner to the next, maintaining and motivated by unsatisfied sexual tension while transitioning among males, egged on in her search for "sustained clitoral stimulation" by the hope that the next guy will finish what the previous one hadn't quite managed. Or maybe if she had already climaxed, she might nonetheless be inspired to encounter multiple males by the simple fact that female orgasm is rekindlable (whether this is the right word for the experience is not certain, but it clearly is a bona fide phenomenon). If so, then the ultimate motivation for such behavior would have been the fitness bonus of taking out an anti-infanticide insurance policy, proximately motivated by the prospect of an orgasm. Or several.

Nonhuman female primates—notably macaques, baboons, and chimpanzees—do in fact mate sequentially with a number of different partners, and if stimulation from these encounters is cumulative, orgasm might be a proximal payoff. It might accordingly be part of a complex reward system among many animals, not just human beings, that induces females to mate with many different males. In addition, it might have evolved in nonhuman primates for one reason (e.g., infanticide insurance), then been maintained among modern human beings for another.

For a distinctly nonsexual example of a trait's adaptive significance changing over time, consider feathers. It is clear from the structure of fossil dinosaurs and birds that the earliest feathers did not evolve in the service of flight because the earliest feathered reptiles lacked hollow bones and were for the most part earthbound. Far more likely, feathers helped primitive "birds" to keep warm and only subsequently became elaborated as aeronautic devices. Among modern birds—especially males—feathers are further adapted as sexually selected traits (remember the peacock). A trait's current adaptive significance can therefore be quite different from its original function.

Even if female orgasm evolved as a mechanism that induced women to mate with multiple men—a questionable if intriguing hypothesis—it doesn't mean that human cultural traditions would necessarily welcome it. Thus, the hideous practice of "female circumcision," still widespread in much of northern and eastern Africa, may owe its existence to a recognition that female sexual desire can lead to multiple partners: in order for a woman to be considered marriageable, it is necessary to guarantee her fidelity by curtailing her orgasmic potential, if not eliminating it altogether. Hrdy has made the interesting proposal that female orgasm may thus be a relic, adaptive among our primate ancestors but potentially disadvantageous—even dangerous—to some women today. Thus, insofar as orgasm might even occasionally induce women to seek out additional sex partners beyond their designated husband, this consequence in itself might have serious (and certainly fitness-reducing) results. In much of the world, the penalty for a woman's having sex with more than one man (especially if she is married) is quite severe, sometimes including death.

As with menstruation, concealed ovulation, the existence of nonlactating breasts, and the other evolutionary enigmas yet to be explored, the conceptual waters surrounding female orgasm are muddy indeed. Thus, another potential evolutionary payoff of orgasm would seem to push in precisely the other direction from Hrdy's hypothesis, toward monogamy rather than multiple partners. Let's return to her suggestion, with the intervening phrases deleted: "It is possible that . . . orgasms have since become secondarily enlisted by humans to serve other ends (such as enhancing pair-bonds)."

There is some evidence that women are more likely to climax with familiar partners because they are more likely to feel (and to be) safe and thus comfortable and relaxed, to be able to make their needs and preferences clear, and more likely to have them met.[23] Put this all together, and a case might be made that rather than being an inducement for polyandry, as Hrdy proposes, female orgasm is an evolutionary sweetener for its opposite, monogamy, as Hrdy also proposes! In this regard, it is altogether consistent to have it both ways because, as already noted, a trait can evolve for one reason, then be employed for another. (In any event, if there were evidence that chimpanzee females are more likely to achieve orgasm by mating sequentially with multiple males, this would strengthen Hrdy's infanticide insurance hypothesis for female orgasm, but diminish confidence in the monogamy hypothesis; at present, there are no data either way.)

A variant on the monogamy hypothesis is also worth considering. Aside from its possible role in promoting pair bonding via the female's inclination, it is not unreasonable to suppose that female orgasm might promote monogamy by acting on the male as follows: if female orgasm serves as a sexual reward, inducing ancestral women to engage in a higher frequency of copulations than would otherwise occur, and if their mates respond positively to an enhanced sex life vis-à-vis that experienced by partners whose females are nonorgasmic, and if part of this response involves a greater commitment to a monogamous pair bond with the women in question, and if this commitment results in greater evolutionary success for these women, then natural selection may have favored female orgasm. That's a lot of "ifs," but—pardon the expression—*if* they all fall into place, so would a case for the validity of this particular hypothesis.

The Monogamy Hypothesis also leads to the prediction that monogamous species should be heavily represented among animals that experience female orgasm, but this is not the case. The best-documented nonhuman examples of female orgasm—bonobos, chimpanzees, certain species of macaque monkeys— are notoriously *non*monogamous.

Perhaps the most persistent hypothesis regarding female orgasm speaks to its most straightforward potential adaptive value: fertilization. Maybe there is some truth after all in the quaint, quasi-Victorian notion, often hinted at but rarely stated explicitly in romance novels, that when a woman truly "gives herself" to her lover, she is more likely to conceive. It would be lovely, if true, not only for its poetic appeal, but because if female orgasm led to greater likeli- hood of fertilization, it would constitute a satisfying book-end match for its male counterpart: he ejaculates when sexually aroused, thereby introducing sperm into the female, and she, if similarly aroused, engages in a kind of internal process that underscores the cooperative or complementary nature of the whole enterprise. Of course, if this correlation were indeed true, then orgasm wouldn't belong in a book about evolutionary enigmas because its adaptive significance would be clear.

But it isn't (clear, that is).

Not that there aren't possibilities by which female orgasm might make fertilization more likely—aside from Desmond Morris's horizontal hypothesis. One theory long favored by many biologists is that the contractions associated with orgasm produce a negative intrauterine pressure, which in turn helps draw sperm up and toward the fallopian tubes. The resulting uterine upsuck hypothesis is linguistically ugly, but logically appealing. Such appeal was enhanced nearly four decades ago by research using a radio telemetry device inserted into the uterus, which in fact showed the production of a vacuum cleaner–like negative pressure following orgasm.[24] Biologists eager (maybe overeager) to embrace an adaptive significance for female climax have cited this study many times—despite the fact that it involved a grand total of one woman!

By-product hypothesis advocate Elisabeth Lloyd points out this egregiously small sample size in her book *The Case of the Female Orgasm*. It is an interesting example of how certain research findings can become reified and embedded in the scientific literature as received wisdom despite the fact that their actual empirical backing may be exceedingly weak. In this regard, sex researchers and evolutionary biologists alike owe a debt of gratitude to Dr. Lloyd, who—a philosopher of science herself—did what is all too rare among practicing empiricists: she actually read the original literature that is so often cited without being critically examined. The article on intrauterine pressure was admittedly in its own way heroic, on the part of the researchers as well as the subject, but it simply cannot (yet?) be accepted as valid. It isn't necessarily *invalid,* but with an "*n* of one," as scientists put it, the case is far from closed. Even worse, follow-on studies attempting to show direct uptake of tiny tracking particles following orgasm have been unsuccessful.[25]

Another possibility presents itself. What if a woman's orgasm increases the likelihood of fertilization by reducing the amount of seminal "flowback"— that awkward but undeniable phenomenon whereby after a man ejaculates, variable amounts of semen leak out of the woman's reproductive tract? Again, at least one study supports this conjecture. A small number of couples surprisingly agreed to collect seminal flowback after sexual intercourse and to report whether female orgasm had occurred and, if so, at what time relative to the man's ejaculation. The key finding was that female orgasm occurring within one minute before and forty-five minutes after ejaculation was associated with less flowback than when there was no orgasm or when orgasm occurred at other times.[26] In other words, it appears that by climaxing, women retain more of their partner's semen.

As with the uterine upsuck hypothesis, the semen retention hypothesis appears plausible, although—once again—its empirical foundation is shaky at best. As Lloyd points out, the data aren't terribly reliable, the sample is small, and the statistical analysis is seriously flawed.[27] And so both hypotheses remain much like female orgasm itself: exciting but elusive.

The (re)search goes on. Female orgasm might facilitate fertilization in other ways. In addition to *possibly* giving sperm a physical boost along their way via negative pressure and keeping them from dribbling out (via reduced "flowback"), it may provide a variety of biochemical benefits, even contribute to the sperm's anatomical development (which is known as *capacitation*). According to researchers David Puts and Khytam Dawood, orgasm may also promote fertilization by

facilitating interaction between sperm and oviductal epithelium, which may prolong sperm longevity, increase the number of capacitated sperm [sperm capable of fertilizing an ovum], or lengthen the interval over which at least some sperm in an ejaculate are capacitated. . . . Female orgasm may allow the earlier entry of sperm into the cervix by resolving the 'vaginal tenting' of sexual arousal, which elevates the cervix from the posterior vaginal wall, removing it from the semen pool. . . . Female orgasm also causes patterns of brain activation and hormone release associated with increased uterine contractions, lower uterine pressure, and movement of semen into the uterus. Peristaltic uterine contractions transport sperm in rats, dogs, cows . . . and probably humans . . . and appear to be caused both by a hormone released during orgasm and by stimulation of brain areas activated during orgasm. In women, orgasm activates the cingulate cortex and medial amygdala . . . , and electrical stimulation of these areas in experimental animals induces uterine contractions. . . . Orgasm also activates the paraventricular nucleus (PVN) . . . , and both PVN stimulation . . . and orgasm have been found to cause oxytocin release into the bloodstream. . . . Oxytocin, in turn, induces uterine contractions . . . , changes uterine pressure from outward to inward, and increases the transport of a semen-like fluid into the uterus and oviducts.[28]

The situation is complex and—at least at first encounter—more than a bit paradoxical. For instance, sperm arrive more quickly at the fallopian tubes in a woman who is *not* sexually aroused! The "vaginal tenting" associated with orgasm raises the cervix away from the posterior wall of the vagina and makes it more difficult for sperm to proceed.[29] As it turns out, however, this delay actually promotes fertilization because freshly deposited sperm cannot penetrate the human egg; they must first undergo capacitation, an anatomical and chemical change that takes one to four hours. Hence, an orgasmically imposed speed bump in the sperm's migration actually makes fertilization *more* likely rather than less.

A useful summary statement comes from a recent scholarly book describing the physiology and anatomy of orgasm (and which, incidentally, was cleverly published with a cover that looks like brown-paper wrapper): "While it is widely recognized that a woman's orgasm is not *essential* to pregnancy . . . [it] may bring into play a combination of physical processes that *promote* pregnancy."[30]

The hormone oxytocin is especially interesting in this regard. Oxytocin is a fascinating chemical involved in inducing uterine contractions associated with giving birth and then helping to induce the postpartum mother to accept and nourish her offspring. It has also been implicated in heterosexual pair bonding in some rodents and might be involved in human pair bonding as well.[31]

For those who agree with the idea that evolution does not play randomly (nonadaptively) with its creations, a modification of Einstein's famous dictum that God does not play dice with the universe, there is fortunately yet another hypothesis regarding the adaptive significance of female orgasm. Better yet, it is not only especially promising, but also encompasses several of the preceding points.

Recall that one of the arguments in support of the by-product hypothesis is that female orgasm is so unreliable that it can't be adaptive. After all, eating is adaptive, and so people get reliably hungry if they are deprived of food. Breathing is adaptive, so their lungs expand and contract reliably. If female orgasm were adaptive, goes the argument, then it too would be predictable. Because it happens only sometimes, it can't be a product of evolution. But wait. This is like saying that because lions do not succeed every time they hunt zebras, it must be nonadaptive for them to hunt! Lots of things are highly adaptive yet don't always work when called on to do so. It is adaptive to find a mate, but not every romantic effort is crowned with success; it is adaptive to reproduce, but some couples—despite their best efforts—are childless. Moreover, one of the most compelling explanations for female orgasm as an adaptation not only is compatible with its elusiveness, but in a sense *relies* on it—more specifically, on the occasional disconnect between body and mind, and how people often find themselves using the former to inform the latter.

The Evaluation Hypothesis

The poet John Donne, writing of one Elizabeth Drury, waxed enthusiastic about how her body not only spoke, but seemed to think:

Her pure, and eloquent blood
Spoke in her cheeks, and so distinctly wrought
That one might almost say, her body thought.[32]

It seems unlikely that any body literally thinks, even the anatomically eloquent Ms. Drury's; that's what brains are for. But if the body does think, then as befits good thought, it does so in silence. And we can't be sure that John Donne, a now-dead white male who wrote several centuries ago, was gesturing toward female orgasm in any case. By contrast, the twentieth-century Portuguese writer and feminist icon Anaïs Nin was less inhibited, writing of "Electric flesh-arrows . . . traversing the body" and noting how "a rainbow of color strikes the eyelids. A foam of music falls over the ears. It is," she announced, "the gong of the orgasm."[33]

With or without an accompanying gong, orgasms sometimes may *appear* to speak, at least to the person who resides inside that body's brain and who might well profit from the information thereby provided.

First suggested by David P. Barash nearly three decades ago,[34] the idea is that orgasm might be a way a woman's body speaks to her brain, "telling herself" that she has been having sex with a suitable partner—that is, one who is not worried about being displaced by a competitor, who is self-confident and unhurried enough to be satisfying to her. When Barash was a graduate student more than ten years earlier, he observed that when subordinate male grizzly bears copulate, their heads are constantly swiveling about on the lookout for a dominant male, who, should he encounter a couple in flagrante, will likely dislodge his lesser rival and take its place. Not surprisingly, subordinate males ejaculate very quickly, whereas dominants take their time. If female grizzly bears were to experience orgasm, with which partner would you expect it to be more likely? And is it surprising that premature ejaculation is a common problem of young, inexperienced men lacking in status and self-confidence? Moreover, is it surprising that women paired with such men are unlikely to be orgasmic?

Research on a large captive group of Japanese macaque monkeys is also suggestive. The technical article's title neatly summarizes its finding: "Female Orgasm Rate Increases with Male Dominance in Japanese Macaques." During 238 hours of observations in which 240 copulations were observed, female orgasmic responses occurred in 80 (33 percent). Of these orgasms, the highest frequency took place when high-ranking males were copulating with low-ranking females, and the lowest between low-ranking males and high-ranking females.[35]

So maybe a woman's orgasm isn't elusive because it is a vestigial by-product, fickle and flaky, sometimes on and sometimes off like a light bulb that isn't firmly screwed into its evolutionary socket. Maybe, instead, it is designed to be more than a little hard to get, adaptive precisely *because* it can't be too readily summoned, so that when it arrives, it means something. The evaluation hypothesis would therefore predict precisely the variability that devotees of the by-product hypothesis consider incompatible with adaptation. It is consistent with several of the hypotheses already described, notably the copulatory reward hypothesis, because it specifies that individuals would be especially rewarded for copulating with partners who are particularly likely to evoke orgasmic responses.

The evaluation hypothesis is even compatible with the fact that orgasm is more reliably evoked by masturbation than by sexual intercourse; potential partners warrant evaluation, whereas there is no comparable pressure to assess one's own masturbatory technique. Moreover, any information made available in the former case can certainly be used to fine-tune the latter. Almost certainly, masturbation is not an adaptation for reproduction in either males or females; rather, it occurs just because the wiring exists—in both sexes—for orgasm based on stimulation, even in the absence of a sexual partner. "If it ain't broke, don't fix it" becomes "if it's available and feels good, use it."

The evaluation hypothesis yields some testable predictions. One, which seems so obvious as to be unworthy of testing, is that women should find orgasms not only pleasurable but important in the context of a sexual relationship. Don't scoff: if a woman's climax is merely an irrelevant, tag-along by-product, then it needn't be accorded any more attention than men give to their nipples. In a survey of 202 Western women of reproductive age, 76 percent reported that experiencing orgasm with a partner was between somewhat important and very important; only 6 percent said it was somewhat unimportant to very unimportant.[36]

If orgasm helps women evaluate their partners, then it would seem to be compatible with an attitude of control and independence. In much of the world, women tend to associate sex with submission, and, interestingly, the more they do so, the more they experience impaired arousability and orgasm frequency, suggesting that orgasms have something to do with autonomy and selfhood, but in an erotic context.[37]

Two more predictions. Compared to their less impressive fellows, socially dominant men should be better lovers—that is, more likely to evoke orgasms in their partners. And for women, experiencing orgasm with a particular partner should lead to preference for that partner. In short, after having had an orgasm, a woman should likely want more and thus have additional sex with the partner in question. The evolutionary outcome is that in the absence of reliable birth control, a woman would more likely be impregnated by this person. Preference for sex with a sexually satisfying lover seems so obvious that it, too, might appear a foregone conclusion, but just because it is obvious doesn't make it any less true or significant. In addition, it is at least possible that causation actually runs the other way: once a woman has a preference for a particular partner (for whatever reason), she might be more likely to be orgasmic with him or her. It might be possible to disentangle these factors, but not easily.

And what about faking orgasm? Many women do it, although the subject has received virtually no research attention beyond the commonsense observation that most men find it gratifying to think that their partner has been "satisfied." Why do men feel this way? The most romantic answer is that if you love someone, you care about his or her happiness, and it isn't genomics science to realize that orgasms make people happy. A more biocynical view—and, not coincidentally, more bioaccurate—would look for an evolutionary payoff for the faker.

Several such payoffs can be imagined. One is that because a woman's orgasm often stimulates a man's, she may be inclined to fake hers just to hurry up his and "get it over with." Or by pretending to be sexually satisfied, a faker might keep her partner more content, self-confident, and thus more likely to maintain the relationship—albeit at the cost of foregoing the kind of intimate feedback that might otherwise make such fakery unnecessary. Alternatively, orgasmic pretense might increase the man's confidence regarding paternity of any offspring, building on his likely assumption that a sexually satisfied woman wouldn't have sought to mate with someone else.

As we have seen, one consequence of being a mammal, with fertilization occurring within the female's body, is that although women can be entirely confident of their genetic relationship to their children, men have to take their wives' word for it. If by being orgasmic—or just seeming to be—women help men move from "maybe" to "definitely," then one might predict that husbands of seemingly orgasmic women would be more likely to invest in their children.

Yet another possible payoff, the most cynical of all, is that faked orgasm is a way of generating a false sense of security, which would diminish the likelihood that the man will engage in "mate guarding," thereby facilitating a woman's ability to engage in extrapair copulations (or EPCs, in the evolutionist lexicon), during which, it is fair to note, her orgasms might even be genuine.

Relatively little is known about actual rates of extrapair copulations, which isn't surprising given the frequently dire consequences of men's discovering that "their" children aren't genetically theirs. Nonetheless, the availability of DNA fingerprinting is rapidly changing this situation, and current estimates are that rates of extrapair paternity are about 2 percent in many human populations and about 10 percent in traditional societies, in which birth control is relatively unavailable and which probably most closely approximate the conditions in which *Homo sapiens* evolved.[38]

Fake orgasm, or something remarkably like it, seems to occur in at least one group of nonhuman animals—certain freshwater salmonid fish that engage in external fertilization. In the best-studied case, that of brown trout, females excavate a depression in a stream bed, whereupon a dominant male typically chases other males away and begins to quiver by vibrating his trunk muscles; subordinate males array themselves nearby. When the female is ready to spawn, she quivers violently, which induces comparable quivering by the male, culminating in the two individuals releasing eggs and sperm, pretty much simultaneously. Sometimes, however, the female quivers but keeps her eggs to herself. Among the various explanations tested to explain this finding, the researchers concluded it was most plausible that female brown trout use "fake orgasm" to trick additional males into joining the spawning pair (her conspicuous quivering attracts additional suitors, some of whom eventually ejaculate). This, in turn, might benefit the female either by increasing the genetic diversity of her offspring or by literally causing a fight to break out among those suitors, after which she gets to spawn—for real, this time—with the winner.[39]

Human orgasms are private affairs (pornographic movies excepted); salmonid fish, by contrast, are altogether shameless. It seems most likely that people use fake orgasms either professionally to titillate viewers of X-rated movies or, on a personal level, to enhance the pair bond or to facilitate a woman's going outside that bond. Among brown trout (which neither produce nor patronize "adult" movies), there is also no pair bond to enhance, so it seems that fake orgasm is strictly a tactic whereby the female gives herself a wider reproductive field from which to choose.

As befits a phenomenon as subtle yet powerful as female orgasm, there is a subtle yet powerful variation on the evaluation hypothesis. What, exactly, is being evaluated? One relevant dimension involves social dominance and self-confidence, à la grizzly bears, but it isn't the only one. Another would include sufficient access to resources to orchestrate interactions that are private, safe, and gratifying—in a word, romantic—and thus appealing to women's evolved evaluation mechanisms. In addition, if orgasms are somehow evaluative, one thing they might evaluate is the partner's genetic quality. A prediction, accordingly, is that female orgasms should be more readily elicited by genetically desirable men.

This correlation does seem to hold. One study has found that women are significantly more orgasmic when paired with men who are more symmetric.[40] Maybe this outcome occurred because of something about the women rather than about the men: Might it be that women who are more orgasmic are differentially drawn to men who are more symmetric? Not so. The same study found that when masturbating, women are equally likely to experience orgasm regardless of how symmetrical their (temporarily absent) partner. Another intriguing result from the same research is that women are more likely to experience ostensibly "high sperm retention orgasms"—that is, climaxes that occurred in close temporal proximity to the man's—when the man is more symmetrical. All these discoveries are consistent with the notion that female orgasm might be a mechanism for "cryptic female sperm choice" by selectively retaining sperm from men who are more genetically desirable or by enhancing the likelihood that a woman will copulate repeatedly with such men, or both. This two-pronged interpretation was first proposed more than twenty years ago and, if anything, now appears more validated than ever.[41]

Here, then, is a prediction: insofar as female orgasm is involved in mate selection by the female—whether by selective retention of sperm or of the guy who produces them—its existence should correlate with species in which females copulate with more than one male. With strict monogamy (as reported for certain South American primates such as the pygmy marmoset) or rigid harem structure (as among gorillas, in which females mate exclusively with the dominant silverback), orgasm should be relatively unimportant. This seems to be the case.

There doesn't appear to have been any comparable research examining the effect of partner symmetry on climax during female-female encounters. It seems reasonable, however, to predict a similar finding: once natural selection has equipped women with a tendency to respond more orgasmically to genetically superior partners—assuming, of course, that it has done so—and once certain individuals find themselves with a same-sex preference (for whatever reason), the likelihood is that they will apply their already-evolved orgasmic inclinations to their preferred partners regardless of gender. It would be helpful if a statistically adequate sample of lesbian lovers would be willing to record their orgasmic histories and if they would team up with researchers equipped with measuring calipers to determine symmetry.

Is it "merely" a Just-So Story to suggest that sexual satisfaction is more frequent and more readily obtained with partners that are more appealing? In a sense, yes. But once a Just-So Story generates predictions and the gathering of evidence—whether the outcome supports or refutes the "story"—it has become a hypothesis being tested scientifically. And the correlation between "sex appeal" and evolutionary payoff is precisely the kind of "obvious" notion that cries out to be evaluated and that sometimes provides gratifying surprises as well as directions for future research.

At the same time, there is more than one dimension to what is "appealing," and of them, body symmetry actually seems rather unimpressive. Admiring a beautiful man or woman, how many people are likely to exclaim, "How wonderfully symmetric"? However, it is entirely possible that an assessment of symmetry figures prominently—albeit unconsciously—in such perceptions. Moreover, symmetry is easy to measure and, so far at least, surprisingly important. Also worth evaluating, however—and perhaps even more important—is the partner's inclination toward the woman.

Male mammals are, in a sense, roving inseminators. Sperm are abundant and cheap, and males, as a result, are primed by evolution to be quick on the draw and not terribly selective as to targets. Their modus operandi is shoot first and ask questions later, if at all. But in certain species, human beings most especially, males have more to contribute: they can be providers, protectors, helpmates, and partners, not just lovers. In addition, a man's behavior as a lover may yield some clues as to his inclination in these other crucial dimensions. According to a simple game theory model, males can be caricatured as either Cads or Dads.[42] Cads are superficially attractive, but lack parental follow-through; they're inclined to love 'em and leave 'em. Dads are, as their name implies, more likely to stay the course and to take the kids to soccer games, but less flashy and perhaps with less instantaneous sex appeal. Might it be that the elusive orgasm has been tuned to help transcend first appearances and encourage women to respond to men who aren't simply out for a quick sexual encounter—that is, to respond in favor of those who are likely to be Dads? If so, how might this work?

If female orgasm were unlocked quickly and easily, then any Cad could do the trick, then be on his way. But, of course, it isn't. Women are somewhat slower to rouse, often requiring extensive foreplay and direct, focused attention to the clitoris, which, after all, isn't within the vagina and thus isn't likely to be stimulated by a hurried and selfish sexual "technique." This requirement, in turn, may have set the stage for a woman to assess whether her partner demonstrates an inclination to be lovingly generous, predisposed to meet her needs rather than selfishly focus only on his own pleasure. If so, then maybe he'll also be inclined to pay for the kids' orthodonture.

"Cryptic choice" of sperm would be especially important in environments in which men tend to provide relatively little investment, in which case orgasm may serve to promote preferential fertilization by "good-quality" sperm or to promote bonding with high-investing men. The Evaluation Hypothesis is thus multipronged and consistent with the fact that nonorgasmic women or women who experience orgasm only rarely are as likely to reproduce as those who are fully and reliably orgasmic. The hypothesis deals with mate selection and ultimately with quality of offspring, not quantity.

Another prediction: if female orgasm is a kind of bioassay of a partner's dadlike tendencies, then there should be a correlation between skill as a lover and inclination or ability to invest as a parent. Someone should test this. Even if such an association exists, however, it is nonetheless possible that women are more sexually responsive to Dads out of gratitude to their daddy-prone behavior. A clearer test, therefore, would be possible, albeit difficult: see whether men whose sexual style involves a notably focused attention to pleasuring their partner turn out, perhaps years later, to be just as focused on taking good care of their children.

Such a finding would actually go counter to one of the widespread assumptions in evolutionary thinking: a woman who perceives her mate to be of low genetic quality may employ a strategy of garnering resources from her primary mate, but at the same time having extrapair sex with a male who is of higher genetic quality. These circumstances could have selected female design favoring retention of sperm from men who possess phenotypic markers of good genes.[43] In other words, women are said to be primed by their biology to settle down with Dads, who offer the long-term, fitness-enhancing prospect of resources and commitment, but to have sex on the sly with Cads, who offer the short-term prospect of sexual excitement, resulting from an equally fitness-enhancing promise of alluring genes to be inherited by their children. Just so. Or maybe not. Whatever the reality, it seems clear that without postulating "stories" of one sort of another, we'll never know the truth. As it is, someday we will.

M enopause—sometimes oddly called *the* menopause—is yet another biological mystery. Less engaging than orgasm, less obvious than breasts, it is in many ways closer to its linguistic partner, menstruation, in being at least semisecret, hormonally hard-wired, and the matching bookend to a woman's reproductive career: beginning with menarche and concluding with menopause. Menopause also shares with menstruation the paradox of seeming to speak of "men" although it occurs only among women (in the word *menopause,* those first four letters are a learned borrowing from the Greek *meno,* "month"). Just as some young women can't wait for the onset of their monthlies, whereas others abhor it, many older women welcome the terminal punctuation mark that puts a period on their periods, but others regret the end of nubility, fear the memento mori, or simply find their attention shifted from uncomfortable menstrual cramps to equally discomfiting hot flashes.

Biologists, too, can't help paying attention to menopause because it is an evolutionary conundrum: Why turn off the possibility of reproduction around age fifty, when most women can still expect several decades of life, much of it quite healthy? After all, evolution is all about breeding. Natural selection is driven by "differential reproduction," which is much simpler than differential calculus or even the differential underneath an automobile. Individuals and genes that reproduce "differentially"—resulting in a positive difference between themselves and their counterparts—will leave more descendants and thus be favored by evolution. So why menopause? Why do perfectly sturdy, happy, competent women consistently lose the capacity to reproduce even though, presum-

6
_____

*The Menopause
Mystery*
_____

ably, their sisters who kept on keeping on would differentially have outreproduced them?

What actually terminates a woman's reproductive capacity? The immediate, proximate answer is well known (as with the other biological puzzles already described): menopause results from a drastic decrease in a woman's endocrine hormones—especially estrogen—correlating with the fact that her ovaries run out of suitable eggs. What remains unknown, once again, is the evolutionary "why." Insofar as natural selection favors maximum breeding success, why do women fetch up against a metaphoric brick wall when it comes to their reproduction?

Women are unusual, perhaps unique, in having a long postreproductive life span. By contrast, men continue making viable sperm throughout their lives, although in diminishing amounts and with gradually decreasing vigor. Age and egg making aren't necessarily incompatible: female centenarian tortoises and even older sturgeons have been reported to breed. Nor is age-caused infertility mandated among mammals: female African elephants, for example, breed into their sixties,[1] and blue whales are still reproductive in their nineties.[2] In fact, most animals remain capable of breeding until they die; by the time a female mammal is no longer fertile, she is almost literally on her last legs. So why do women, unlike nearly every other species (with the currently inexplicable exception of the short-finned pilot whale[3]), lose the ability to reproduce when they still have decades to go?[4]

New Old Age?

One possibility is that menopause is a recent arrival on the human scene. Or perhaps the aberration isn't menopause, but the fact that people in general and women in particular are living longer than ever before. Certainly, ancestral human populations didn't have the benefits of public hygiene, childhood immunizations, antibiotics, and abundant food, so maybe our Pleistocene biology was pretuned to dying by age fifty or so.

Here, then, is another nonadaptive hypothesis, the new old-age hypothesis, that deserves attention, but as with the by-product hypothesis for female orgasm, it doesn't do well under scrutiny. (We make this statement with some regret because we would love to embrace at least one bona fide nonadaptive explanation if only to demonstrate—or feign—our own admirable open-mindedness. Science, however, is a matter of plausibility, testability, and intellectual fertility, not personal preference.)

Here's why the new old-age hypothesis is unconvincing.

For one thing, even a small selection differential can drive substantial evolutionary changes, and it would seem that in the distant past there were plenty of women who lived into their sixth, seventh, and eighth decades; if even a small proportion of them continued to reproduce, their additional fitness would have selected against menopause. For another, it isn't at all clear that when it comes to the human species, "old" is all that new. Part of the problem is with the data. Although there is little difficulty distinguishing a fossil adolescent's skeleton from an adult's, there are no reliable ways to determine whether ancient remains are from, say, a forty-year-old or a sixty-year-old. As a result, the presumption that prehistoric life spans were necessarily shorter than today's remains just that, a presumption.

An additional problem comes from misreading statistics. Mark Twain once identified three kinds of lies: lies, damned lies, and statistics. Let's imagine, for example, that out of four individuals who died on the Pleistocene savannah, two lived to be eighty years of age, whereas the other two died in early infancy. Statistically speaking, the average life span was therefore forty (2 times 80 plus 2 times 0 = 160, divided by 4 = 40), but this does not mean that a fifty-year-old would have been unusual or even considered especially elderly. In our hypothetical sample, half the population was going to live an additional thirty years! Similarly, although life expectancy *at birth* among contemporary hunter-gatherers is only about thirty to thirty-five years, this is due to high infant and juvenile mortalities.

Most of the statistically significant, increased human life span that characterizes the twenty-first century compared with the life span of our Pleistocene forebears is due to reduced juvenile mortality. The maximum ceiling on human age hasn't changed in centuries, probably millennia. Among hunter-gatherers today, once someone has survived infancy and childhood, the likelihood is that she will survive at least into her fifties and often to a greater age. Life expectancy *at age twenty* for a contemporary forager is about twenty years beyond the age of menopause. Having reached sixty, she will probably live to be more than seventy. So the oft-cited fact that the "average life span" in prehistory was something like thirty-five years doesn't mean that thirty-eight-year-olds were rare; rather, a large number of early deaths skews that average. (For another take on the misleading nature of statistical averages, consider that the average human being alive today has precisely one ovary and one testicle.)

We can be assured that menopause has been around for at least three millennia and probably much longer. Frequent cross-cultural references to our allotted life span, such as the biblical "three score and ten," suggest that although the elderly may have been less abundant than today, they weren't all that anomalous. Thus, Genesis (18:11) speaks of the aged Sarah: "It ceased to be with Sarah after the manner of women," suggesting not only that Abraham's wife had stopped menstruating, but that such a circumstance was well known to the author(s). Becoming predictably infertile was, as it is now, "the manner of women." Moreover, menopause doesn't happen just to women who enjoy the blessings of modern public health and thus live a long time: whether hunter-gatherers of Africa, South America, or New Guinea; the Inuit of Alaska; nomadic herders of Mongolia; pampered, highly nourished, gym-frequenting, carefully doctored debutantes; or Manhattan society matrons, *all* women become nonfertile within a few years of age fifty. Not only that, but modern techniques of artificial fertilization have made it possible for women in their fifties and even sixties to sustain pregnancy and bear healthy babies, showing that even an aging female body can be up to the task, which makes it all the more paradoxical that the woman's reproductive organs will have shut down previously.

Another strike against the new old-age hypothesis for menopause—if one were needed—is the curious fact that it involves the reproductive system only. If menopause occurs because modern developments have kept the human body going beyond its usual and allotted life span, why don't we run out of gas in other ways? Why don't human kidneys consistently give up in the sixth decade? Why doesn't the liver or the heart or the brain throw in the towel? Evolutionary theory strongly suggests that when it comes to the debility of old age, all systems would fall apart at about the same time because if any one began to fail early, selection would no longer operate to maintain the others.

The story is told that Henry Ford once commissioned some of his engineers to examine Model Ts in junkyards, looking for which parts wore out and which were still usable. It was discovered that the kingpins remained consistently healthy, even when the pistons, rings, drive shafts, and so forth showed substantial wear. Mr. Ford then ordered that Model T kingpins be built to *lower* specifications.

To some extent, in fact, living things are engineered similarly, such that elderly bodies tend to wear out from the simultaneous failure of any number of organ systems; none of them—the lungs, the heart, the liver, the kidneys—is notably over- or underbuilt. In Oliver Wendell Holmes's delightful poem "The Deacon's Masterpiece, or the Wonderful One-Hoss Shay," subtitled "A Logical Story," we learn of a deacon's determination to build the perfect horse-drawn carriage, which he constructed using only the highest-quality material for every part. As a result, there was no weakest point, and so the wonderful one-hoss shay never broke down until one day there was

> The poor old chaise in a heap or mound,
> As if it had been to the mill and ground!
> You see, of course, if you 're not a dunce,
> How it went to pieces all at once,—
> All at once, and nothing first,—
> Just as bubbles do when they burst.
> End of the wonderful one-hoss-shay.
> Logic is logic. That's all I say.[5]

People are pretty much constructed (have evolved) like the Deacon's Masterpiece, with all systems about equally durable—except for women's reproductive system, which stands out as a notable exception. It is as though the wonderful one-hoss shay was built with papier-mâché wheels that stopped working long before any other component. Logic is logic, as Justice Holmes points out, and it would be illogical in the extreme if evolution worked this way. Far more likely would be a scenario in which there is something special about women's reproduction so that, for some reason, evolution targeted it for termination. Yet, ironically, natural selection is precisely the process that promotes reproduction and that should therefore support it above all other things. An enigma indeed.

It is not merely reproduction that is in the biological cross-hairs, but *female* reproduction. Despite the occasional talk about "male menopause," the fact is that the latter doesn't really exist, or at least in nothing remotely like its female counterpart. Men into their eighties and, according to some accounts, even their nineties have fathered children. Men's fertility declines gradually, at about the same pace as the aging of their other organ systems, whereas women's fertility stops abruptly. If "reproductive senescence" is a result of an artificially extended life span, why does it single out (1) reproduction and (2) *women's* reproduction?

Out of Eggs?

At this point, another nonadaptive hypothesis rears its head, although it doesn't appear to deserve serious consideration. Menopause occurs when a woman runs out of developing follicles, so perhaps her reproductive system shuts down when there is no more reproducing left to do. At first glance, this out-of-eggs hypothesis seems plausible, maybe even a slam-dunk. You can't drive an automobile when the tank is empty, and, similarly, a woman can't very well ovulate when she's out of eggs. But this "explanation" has no real explanatory power because it confuses proximate with ultimate causation. The "ultimate" reason for running out of gas *isn't* that a car can't run when it's out of gas, but that someone didn't fill the tank. With gas, a car runs. The "ultimate" reason for menopause isn't running out of eggs, but not having been provided with more. So another question is, Why don't women have more eggs?

Every female fetus has about 7 million primordial germ cells within her developing ovaries. These cells are whittled down to only 400 or so ovulated eggs: roughly one each month (a dozen per year) over perhaps thirty-five potentially reproductive years. There seems no reason why 600 competent eggs wouldn't be recruited if there were an evolutionary payoff to reproducing over, say, a sixty-year span. If it were somehow necessary to sort through 150,000 ova just to find one worth developing, then can't an immature ovary be preloaded with 10 million early germ cells instead of a "mere" 7 million?

Moreover, think about those modern-day women who do not practice birth control and who therefore spend much of their reproductive lives either pregnant or lactating and therefore release only one-half or one-third of their lifetime egg supply: they nevertheless enter menopause around age fifty just like everyone else, despite having all those unused eggs and no later than their sisters who supposedly become menopausal because they used up all of theirs.

The conclusion is unavoidable. Menopause is not an illness or an incidental result of modernity. Nor is it testimony to evolution's lack of foresight in which natural selection forgot to fill 'er up with enough follicles. To be sure, menopause is a genuine inconvenience for many women because its associated hormonal changes interrupt sleep and can be unpleasant, even temporarily disabling. In theory, it is conceivable—although not very—that the hot flashes and other correlates of menopause aren't correlates at all, but rather the main event: that selection has favored unpredictable, unpleasant sensations of sudden warmth in middle-aged women and that ceasing ovulation is itself a by-product of these sensations. Don't expect anyone, however, to take this explanation seriously. Hormonal fluctuations and its consequences have almost certainly not been selected for. They at least are genuine by-products of the main event: the fact that menopause has been selected for in women.

The Patriarch Hypothesis

But before we leave the world of by-products altogether, we should look at yet another nonadaptive possibility, akin in its androcentrism to the by-product hypothesis already described for female orgasm and equally far-fetched. Maybe it isn't that women have been selected to stop reproducing at a certain point in their lives, but rather that selection resulted in their living longer because extended life span was favored in *men,* and women were simply unintended beneficiaries. We have already described *Homo sapiens'* biological predisposition for polygyny, such that a small number of successful men have prehistorically garnered more than their share of reproductive success.

The idea is that once human sociality reached a level of complexity such that a few highly successful males were able to dominate the breeding fray into middle age and beyond, selection would favor extending the lives of these men until they fell apart for a variety of reasons (heart disease, cancer, etc.). Dubbed the patriarch hypothesis,[6] this notion suggests that if the relevant genes for extended life span were located otherwise than on the Y chromosome, women too would have experienced the effect, being "dragged along" by men. (The Y chromosome is passed on from father to son and is not present in women; hence, so long as any such longevity factors occurred on other chromosomes, they would be inherited by both sexes, even if directly selected for only in men.)

The patriarch hypothesis is ingenious but unsatisfying because it seems likely that once women enjoyed an extended life span—for whatever reason—selection should have increased their allotment of ova so that they would be released as long as a woman is alive, if there were an adaptive payoff to doing so. Thus, the hypothesis falls victim to the same problem as the out-of-eggs hypothesis. If the patriarch hypothesis is correct, and men were selected to live longer because at least some of them were able to keep breeding, which in turn selected for continued sperm production, why weren't women similarly selected for continued egg production when as a by-product they, too, started living longer? Why pause for menopause?

Let's put the focus where it belongs, on women.

Women get pregnant; men don't. Women give birth; men don't. Women lactate; men don't. And women pay a biological price for all of these abilities, which men don't. *Maternal* mortality during childbirth can be deplorably high, especially if, for example, the baby's head is too large for the mother's birth canal (a problem called *cephalo-pelvic disproportion*) and safe Cesarean procedures aren't available. *Paternal* mortality during childbirth? Zero. If we add to this fact the regrettable reality of aging—the fact that elderly bodies are less viable and more breakable than their younger counterparts—it seems likely that bearing children exerts an increasing toll with increasing age and that however great this toll, it is borne by women rather than by men. Small wonder, then, that at some point in their lives women stop bearing children.

It isn't that simple, however. Even though basic biology would seem to dictate reproductive cessation, there remain a few details to be worked out. For example, as with orgasm, natural selection wouldn't keep middle-aged women from reproducing simply as a gift of benevolent infertility, to keep them alive, like a governor installed on an automobile, intended to keep the driver safe by restricting his or her speed. Staying alive is often a human hope, consciously striven for, but it isn't an evolutionary goal, at least not unto itself. Natural selection wouldn't install menopause in human beings as a pregnancy-inhibiting, life-preserving governor unless those who were thereby "governed" ended up producing more offspring—or, more properly, projecting more copies of menopause-promoting genes—than would otherwise happen. It wouldn't matter to evolution if women kept on getting pregnant and died in the process; indeed, unrestricted, age-indifferent breeding unto the bitter end is the case for nearly every other animal species. It must be the case that for some reason women who inhibited their reproduction in middle age didn't merely live longer, but actually produced *more* genetic descendants.

It is easy to see how evolution would favor extended life spans so long as those at the upper end of the span are reproducing: because a thirty-year-old, for example, would leave more genetic descendants than a twenty-year-old, selection would favor living at least to age thirty. But if fertility terminates abruptly and permanently by one's early fifties, why should evolution favor a life span of seventy or more years instead of sixty?

Time for some more hypotheses.

The Prudent Mother

Which came first, the chicken (postreproductive life span—women living for decades after they have stopped ovulating) or the egg (menopause, as such— that is, an end to egg release in middle age)? Did the evolution of menopause occur because selection extended the female life span beyond the point when senescence would otherwise take over or because it directly favored an early termination of female fertility, preventing ovulation and pregnancy when many women are still quite healthy?

Most biologists favor the latter explanation, and for good reason. Whatever the cause, "early senescence" is an anomaly in the living world not only because evolution only rarely favors not breeding, but also because in most respects human evolution has involved selection for aging that is *delayed* rather than premature. Not only do human beings typically live considerably longer than other mammals, but we are Peter Pans among primates, wherein evolution promoted such traits as hairlessness, large heads, and an overall degree of neotony, or extended childhood combined with delayed development. Given that evolution has apparently been telling our ancestors to slow down, take their time, grow and mature slowly, and even to some extent remain infantile in appearance even when sexually mature, it is more than a bit surprising that it turns around and favors reproductive restraint just when things have finally gotten rolling.

Perhaps it's a matter of prudence.

Case in point: Flo's failure. This female chimp, well known to those who have followed Jane Goodall's detailed studies of the behavior of free-living chimpanzees, was an iconic matriarch who—like other nonhuman primates, but, if anything, somewhat more so—kept on breeding into advanced old age. Indeed, it seemed a near miracle when Flo became pregnant for the last time because she was obviously dilapidated in every respect. But she must have dipped into the reproductive well one time too often. Here is the sad story, reported by the Jane Goodall Institute:

Flo gave birth to at least five offspring: Faben, Figan, Fifi, Flint, and Flame. She was a wonderful, supportive, affectionate and playful mother to the first three. But she looked very old when the time came to wean young Flint, and she had not fully succeeded in weaning him when she gave birth to Flame. By this time she seemed exhausted and unable to cope with the aggressive demands and tantrums of Flint, who wanted to ride on her back and sleep with her even after the birth of his new sister. She still had not weaned Flint when Flame died at the age of six months, and at this point stopped even trying to push Flint to independence. Flint therefore became abnormally dependent on his old mother. When Flo died in 1972, he was unable to cope without her. He stopped eating and interacting with others and showed signs of clinical depression. Soon thereafter, Flint's immune system became too weak to keep him alive. He died at the age of eight and a half, within one month of losing his mother Flo.[7]

It is unfair to call Flo a failure; she had at least three flourishing offspring and was by every measure a maternal and evolutionary success. But it is also hard to avoid the conclusion that Flame, her last and youngest child, died because Flo at age forty-three or so was simply too old. In addition to possibly inheriting some invisible genetic defect (after all, the older the mother, the greater the prospect of genetically impaired offspring), Flame may have received insufficient or inadequate nutrition in utero or during lactation. Almost certainly, Flo wasn't up to the task of bringing one more baby chimp into the world. Not only that, but the effort may well have hastened her own demise, *and* had she refrained, perhaps she would have been able to take better care of Flint, who might then have survived. Instead of Flo as failure, perhaps we should therefore identify Flo's folly: not knowing when to stop attempting to breed yet one more time when restraint would have yielded higher overall fitness.

Meditating on this melancholy episode, Sarah Hrdy has proposed the prudent mother hypothesis by which menopause is essentially a way that women avoid Flo's folly. "Stopping early," suggests Hrdy, "guarantees that old mothers who give birth to one last baby will care for it long enough to see this reproductive enterprise through to a successful conclusion."[8]

It is a compelling argument. After all, human beings, largely because of their insistence on walking upright, which effectively narrows the birth canal, face a genuine risk of dying in childbirth, one that is insignificant in most other animals. Jared Diamond notes that

in one study encompassing 401 pregnant female rhesus macaques, only one died in childbirth. For humans in traditional societies, the risk was much higher and increased with age. Even in affluent, twentieth-century Western societies, the risk of dying in childbirth is seven times higher for a mother over the age of forty than for a twenty-year-old mother. But each new child puts the mother's life at risk not only because of the immediate risk of death in childbirth but also because of the delayed risk of death related to exhaustion by lactation, carrying a young child, and working to feed more mouths.[9]

Add to these risks the fact that mothers are obviously key to children's survival. Thus, a review of twenty-eight different "natural fertility, natural mortality populations" conducted to examine the association between death of mother and death of child found that child mortality is directly keyed to maternal survival.[10] This effect declines rapidly as the child ages, being most in-

Flo, a celebrated but possibly "imprudent" chimpanzee mother, carrying one of her offspring (most likely Flint). Photograph by Hugo Van Lawick; used by permission of the Jane Goodall Institute, www.janegoodall.org.

tense—not surprisingly—when a mother dies before her child is weaned. But things aren't quite so simple. For one thing, although there is no question that older women are at significantly greater risk of producing a genetically defective child, there is still debate about whether older women really are at greater risk of dying during childbirth than are younger ones, at least among those nontechnological, "natural fertility, natural mortality populations" so beloved of anthropologists.[11]

Most human beings also do not produce a follow-on baby until the current one is weaned, which was presumably true especially in the nontechnological environment in which *Homo sapiens* evolved. As a result, the cost to a weaned child of losing its mother—although doubtless substantial—would have been less severe than had that parent died during lactation. Moreover, Flo's folly is easy to label in retrospect after she died (along with Flame and the unfortunate Flint); had she bred successfully yet again, her numerous fans would instead have celebrated Flo's fertility—and, more to the point, so would natural selection, which would have rewarded her with even higher genetic representation in the next generation. There would have been yet more fertile Flo genes in the population, increasing the likelihood that future chimps would go with the Flo and keep breeding into advanced old age.

Standard life-history theory—a cornerstone of modern population biology, well supported by mathematical analyses—suggests that "reproductive effort" should increase with age, and by and large it does. An older bird, unlikely to breed again and thus literally having all her eggs in one last basket, will typically face greater risks defending those eggs against a predator; by contrast, younger ones—those who are capable of breeding another day—are more prone to run away. Standard procedure at many hospitals is that when a maternity patient is identified as an "older primip" (from the term *primiparous,* which means "having offspring for the first time"), there is often a bias in favor of riskier procedures if needed to deliver a healthy baby.

In the poker game of breeding in which maximizing your fitness substitutes for maximizing your pile of chips, just as there is a payoff for betting successfully on one's breeding potential, there is a cost to betting too low, thereby underplaying one's hand. As in Kenny Rogers's song "The Gambler," "you got to know when to hold 'em, / know when to fold 'em, / know when to walk away / and know when to run."[12] The Prudent Mother Hypothesis is that menopause tells women when to fold 'em. But it probably isn't the only game in town.

Good Grandmothers, Part I

Science tries to explain things, and if we wait long enough, it nearly always succeeds. Along the way—and not merely incidentally, but *in order to* make progress—it poses hypotheses about why things are as they are: in other words, Just-So Stories, which, if they serve more than merely an entertainment function, lead to the scientific testing of specific hypotheses. A not uncommon frustration occurs en route, however, when multiple hypotheses point in the same direction, such that more than one can account for a given outcome. This has been the case for many of the evolutionary mysteries in women that we have encountered thus far, and menopause is no exception.

Neither lions nor baboons are normally thought to undergo menopause. Yet female baboons in the Gombe Wildlife Reserve—home to Flo and those other chimps so famously studied by Jane Goodall—live to a maximum of twenty-seven years although their fertility drops dramatically after age thirteen. Similarly, lionesses in Serengeti National Park live up to seventeen years although their fertility declines after age thirteen.[13]

Old lionesses and baboonesses live for only a few years after they stop reproducing, long enough just to ensure that their last-born offspring receive the mothering they need. Thus, they qualify as prudent mothers, their prudence manifested via reproductive restraint. As noted, however, many women typically flourish for literally decades in a nonovulating, nonbreeding state, perhaps because human childhood dependency is so much longer and because (associated with our species' prolongation of childhood) a reproducing woman typically finds herself simultaneously caring for several dependant offspring of different ages.

Thus, even as the prudent mother hypothesis makes a great deal of sense, it seems only prudent to acknowledge that menopause doesn't just help mothers stay alive longer, even if by doing so they can give a boost to their youngest, last-blooming children. It may also help them to produce more successful grandchildren. Another way of framing this idea—a way that will make more sense a bit later—is that success of one's children and thereby of one's self occurs only when those children are themselves successful. And so welcome to the grandmother hypothesis.

Perhaps it is merely coincidence, but just at the time when a woman undergoes menopause, her younger children are becoming parents in their own right. Around the world, grandmothers pitch in and help out. Perhaps it is mere coincidence once again, but those who do so typically end up with more grandchildren than those who don't. Hence, the hypothesis that women stop reproducing when they do because by doing so they are freed up to be doting grandmothers and that given the increasing downsides to bearing children as age increases, there is a greater payoff to grandmothering than in trying to be a mother one more time.

Note that being a prudent mother doesn't preclude being a beneficent grandmother. In fact, in a species such as *Homo sapiens,* in which the generations overlap, the former would seem to lead almost inevitably to the latter, making it very difficult to unravel the two. Once you are no longer encumbered with children of your own and especially if your youngest offspring is weaned and merrily on his or her way, it makes perfect sense—socially as well as biologically—to pitch in and help with your kids' kids.

The implications are intriguing. It is a challenge to explain why, among natural fertility populations, human beings actually have shorter interbirth intervals than do other great apes. By all accounts, human babies require more parental care and attention than do any other mammals, yet in the absence of birth-control methodology women give birth about every three years, compared to every four to five years for chimpanzees and approximately every eight years for orangutans.[14] How can this be?

One possibility, more like a probability, is that something as basic as the biology of birth spacing evolved in the context of something equally basic—namely, the presence of others who helped out. And what is more basic than a close relative—in particular the mother's mother? Either she is nonreproductive because she has been rendered prudent by selection's operating on her via the direct benefits she can bestow on her offspring, most likely her most recent child, or her postmenopausal status has arrived specifically because she can take up some of the maternal obligations that would otherwise descend on her somewhat older children, the ones who are making her a grandmother.

Consider the Hadza, hunter-gatherers living in modern-day Tanzania, but whose lifestyles hearken to something resembling the conditions under which *Homo sapiens* evolved. There aren't many Hadza left (about a thousand), and they are increasingly pressured to take up a sedentary, "modern" way of life. But the Hadza are stubborn, and for the most part they disdain agriculture and herding. Hadza men hunt, women gather and forage, and the most energetic and productive foragers of all are—you guessed it—postmenopausal women.

Elderly Hadza woman with digging stick. Photograph from AnthroPhoto.

Extensive studies by anthropologist Kristen Hawkes and her colleagues have documented that Hadza grandmothers work harder than anyone else, providing the lion's share of the calories, mostly by digging for roots, gathering fruit, and bringing home impressive quantities of honey.[15] Younger adults, who are presumably healthier and more agile, are also lazier or otherwise occupied: teenagers and newlyweds forage for a scant three hours per day; married women with young children are out digging and gathering for about four and a half hours per day. And grandmothers? A full seven hours. Moreover, in terms of calories brought back per hour of work, these elderly, hard-working grannies are no less efficient than are women in their twenties and thirties. As a result, they bring back more food, far more than they themselves need, and they give this food to their children and grandchildren, whose body weight varies directly with their grandmothers' food-gathering efforts. For a society in which infants are carried around rather than left in a crib or stroller, this gift of food isn't mere token support. A mother, no matter how healthy, hardy, and determined, is necessarily constrained while burdened with a baby. It is doubtless significant than among the Hadza, *every nursing mother has a post-menopausal helper.*

Nor is provisioning the only benefit grandmothers dole out: they also baby-sit, step in to mediate disputes, lend their social status to their younger family members, and so forth. In fact, the ubiquity of assistants, notably although not exclusively postmenopausal grannies, has led a growing contingent of anthropologists and evolutionary biologists to conclude that the so-called nuclear family is not fundamental to the evolutionary history of *Homo sapiens*; rather, human beings are "naturally" cooperative breeders.[16] Cooperative breeding is well known for a number of bird species, which make use of so-called helpers at the nest, usually older offspring from earlier breeding efforts, who help provision the next brood, defend the nest site, even sometimes risk their lives to drive away a predator. Such cooperation is less common among mammals, if only because female mammals are uniquely adapted to nourish their own offspring (don't forget those lactating breasts). But it does occur on occasion—for example, among African meerkats,[17] whose exploits have become well-known to a generation of television fans via *Meerkat Manor.*

The technical term for cooperative breeding is *alloparenting,* from *allo,* meaning "other." It is a concept that gives pause, especially in present-day America, which generally devalues cooperation in favor of autonomy. Many common English words—*autonomy* not least—begin with *auto,* for "self," such as *automobile* (self-moving vehicle) and *automatic* (self-activating whatever);

Meerkats congregating outside their burrows. Photograph by Fotosearch.

not only are there no *allomobiles* or *allomatic* devices, but it requires a dictionary to find any *allo*-anythings. Yet alloparenting may well have been crucial to human evolution, with an especially vigorous push from and reliance on grandparents, in particular grandmothers.

If so, then one cannot help wondering about the extent to which modern technological life, with its emphasis on geographic mobility and its disparagement of adult children's rearing their offspring with alloparental assistance, might be a source of considerable stress. It is almost certainly an evolutionary novelty for young adult human beings to rear their own offspring without the alloparental assistance with which *Homo sapiens* almost certainly evolved. Anthropologists use the term *matrilocal* to indicate social systems in which a newly married couple is expected to reside with or near the bride's family; *patrilocal* means that the bride lives with or near the groom's family; and *neolocal* characterizes the most common situation in North America, wherein a young couple is expected to set up separate housekeeping independent of either the bride's or the groom's family. A more appropriate term for the latter situation might be *nullilocal*, near nobody, which emphasizes the absence of social and biological support that frequently results.

The likely cost is probably also borne by young children, who struggle to grow up with a dearth of alloparents, and of course by those grandparents (postmenopausal grandmothers in particular), who may well owe their own postmenopausal state to the prospect—these days often negated, especially in the highly technologized West—of providing such alloparental assistance.

Grandfathers? (A Brief Digression)

Beyond the question of alloparents, especially grandmothers, what about those other direct parents—fathers? And what about grandfathers?

The reality is somewhat discouraging. Although *Homo sapiens* are more paternal than any other primate, the truth is that in no human society do fathers do as much fathering as mothers do mothering. Any cross-cultural universal of this sort cries out for a biological explanation because it is inconceivable that random, arbitrary social tradition would have hit independently upon the same arrangement (mothering is greater than fathering) in every case. Most likely, universal traits are attributable to something else that is shared, and all people, from Greenland to the Gobi, regardless of their differing cultural traditions, share one thing—their biology.

In this case, the key cross-cultural template seems to be confidence of genetic relatedness—not conscious, self-aware confidence, but rather a consistent evolutionary pattern, reliable enough to have driven natural selection in a particular direction. It is surely no coincidence, for example, that out of about four thousand different species of mammals, there is not one in which the males lactate,[18] even though doing so would seem only fair, even perhaps adaptive, given that the female has recently undergone the rigors of pregnancy and childbirth. But because of females' internal fertilization, male mammals (including men) cannot have complete confidence that they are related to "their" offspring; female mammals (including women) can.

Although men don't lactate, they do contribute to their offspring—just not as much as women do. They may be important in defending their families, providing other forms of parental assistance, and so forth. They also typically provide food as well, but like male chimpanzees who use meat obtained through hunting to cement relationships with additional females, men are more likely than women to share the bounty of extra calories with nonrelatives—not because they are more generous, but because compared with women, they are more likely to enhance their fitness by using surplus resources to enhance their status and perhaps secure additional copulations. (It is rare indeed for female mammals to have to resort to food bribes in order to obtain a mating.) Besides, unlike women, men cannot be certain that resources donated to a child are going to *their* child and not to someone else's.

What about grandfathering? Surprisingly little information has been gathered on the evolutionary payoffs to men of living into their advanced years, probably because men remain physiologically capable of producing sperm, so there is no dramatic reproductive punctuation mark like menopause in women that cries out to be explained. In strictly monogamous societies, however, there should be little biological benefit to prolonging a man's life span beyond that of his wife's reproductive years, unless grandfathers somehow contribute to their own fitness. Detailed study of church-recorded birth, death, and marriage records in premodern Finland showed no such benefit: although postreproductive women gained on average two extra grandchildren for every ten years that they survived beyond age fifty, the presence of grandfathers was not associated with their children's producing additional offspring of their own—that is, no extra grandchildren for the grandfathers.[19]

As a result, we are faced with the curious problem that extended life span for *men* becomes a paradox, too, perhaps even more so than the existence of menopause in women! Several explanations are available, none clearly correct.

Maybe extrapair matings have been more frequent, even in some presumably monogamous societies, than generally supposed. Or maybe such matings were especially frequent in the distant hominid past, and although cultural practices have changed since then, human biology hasn't. Or perhaps the earlier-described patriarch hypothesis is precisely backward, and men's longevity is actually a by-product of selection's having favored postreproductive grandmothering on the part of women!

In any event, the situation is notably unclear because few researchers have bothered to look for the reproductive costs or benefits of grandfathering, although the basic pattern just described for premodern Finland also seems to occur in at least one Caribbean population.[20] By contrast, evidence from premodern Finland and Canada shows that when postreproductive mothers were around, their adult offspring bred earlier, more often, and more successfully. Not only that, but the fitness benefits derived by these grandmothers declined as the parental success of their own offspring declined—which is just the point at which elderly grandmothers began to die off.[21]

One might predict that although cross-generational maternal links remain strong (because confidence of genetic connection doesn't diminish in the female line), comparable paternal links would reduce the tendency for parental solicitude on their part. Unlike a mother, each father is in a sense a genetic weak link, at least when it comes to reliable genetic connectedness. Every child has four grandparents: mother's mother (MoMo), mother's father (MoFa), father's mother (FaMo), and father's father (FaFa). MoMo is guaranteed seamless genetic continuity; MoFa and FaMo have one link of paternal uncertainty; FaFa has the lowest genetic confidence because he has two such weak links. The following prediction can therefore be made: MoMos should be more helpfully grandparental than either MoFas or FaMos, whereas FaFas should be the least grandparental of all.

A survey of 120 American undergraduates found that they consistently ranked their grandparents in precisely this order with regard to "closeness," time spent together, and—one readily measured factor—grandparents' financial contributions: FaFas were ranked last, MoFas and FaMos were intermediate and not statistically distinguishable, and MoMos were considered the most solicitous.[22] This same pattern of "discriminative grandparental solicitude" was independently confirmed in a study of German students, further suggesting that it isn't merely a cultural artifact, but a consequence of biological proddings.[23]

There is even some evidence that the presence of grandfathers sometimes actually reduced the well-being of their grandchildren, possibly because the

EVOLUTIONARY HYPOTHESES EXPLAINING MENOPAUSE

New old-age: recently extended lifespan
Out of eggs: nonadaptive
Patriarch: a by-product of male lifespan
Prudent mother: self-preservation
Grandmother 1: alloparenting
Grandmother 2: inclusive fitness
Competition avoidance: giving way to the next generation
Female longevity 1: backup chromosomes
Female longevity 2: sexual selection on males
Female longevity 3: "grandmothering"

grandfathers received preferential access to food.[24] In such cases, males' ability to reproduce throughout their life span, combined with diminished confidence of relatedness to their children vis-à-vis that of females, may have selected for a greater inclination to invest in themselves and their immediate offspring. One thing seems clear: despite considerable diversity from one human group to another, grandfathering is generally less likely to pay evolutionary dividends than is grandmothering, and this finding, in turn, is consistent with the unquestionably biological, cross-cultural fact that menopause is definite among women once they reach a "certain age."

Don't give up on Heidi's grandfather, however. At least one study failed to confirm grandmotherly bias, this time in rural Greece.[25] Nor is this finding entirely surprising, considering that, after all, both biology and culture are crucial, complex determinants of something so elaborate as grandparenting, and neither can entirely override the other. Imagine, for example, that a society suddenly conferred great rewards on people with an eye in the middle of their forehead; despite the most enthusiastic cultural encouragement, it is unlikely that large numbers of little Cyclopses would accordingly be born. Similarly, it is not off-putting to evolutionists (and is perhaps even somewhat reassuring for anyone committed to cultural diversity) that despite biology's likely whisperings, male-biased grandparental involvement remains the norm in a rigidly patriarchal and patrilocal society such as rural Greece.

The specieswide norm, however, is otherwise. Thus, a study of births, deaths, and family residence patterns in a village in central Japan between 1671 and 1871, which looked for effects of presence or absence of various family members on the probability of a child's death, found that the only grandparent whose presence was consistently associated with reduced childhood mortality was the mother's mother.[26] In addition, maternal grandmothers' (MoMos') positive contribution seems to extend across numerous cultures, disconnected from each other. A typical research report, for example, was titled, "Maternal Grandmothers Improve the Nutritional Status and Survival of Children in Rural Gambia."[27]

Back to Grandmothers

Alloparenting (especially by grandmothers) isn't limited to human beings or telegenic meerkats. A study examining captive vervet monkeys found that infants are more likely to survive if their social group includes at least one grandmother. In addition, removing the mother of a healthy, breeding female caused a decline in the female's breeding—the interbirth interval increased, and those infants that were produced were less likely to survive.[28] Perhaps this outcome is a consequence of captivity, but probably not. Among free-living baboons, for example, having a grandmother in the group also results in higher survivorship among babies.[29] In both the vervet and baboon cases, Grandma provides social support rather than food: after baboons and vervets are weaned, they—like nearly all other primates—get their food by individual foraging. Unlike people, they don't bring food back and then share it. Of course, neither elderly vervets nor elderly baboons are typically postmenopausal because they don't go through menopause at all, and the beneficial effect of their presence, moreover, is relatively modest.

But that finding is precisely to be expected if menopause in humans has been selected because of benefits provided by nonreproductive women to their grandchildren: because elderly vervet and baboon matriarchs can't contribute very much and their mortality during childbirth is far lower than that of *Homo sapiens,* they don't experience anything comparable to human menopause. It seems a good bet that if tomorrow someone discovers a species of nonhuman primate that undergoes menopause, that species will also be one in which postmenopausal individuals are generous, hard-working, baby-promoting grandmothers.

Sarah Hrdy tells of a particular langur monkey, "old Sol," who had ceased menstruating (and thus might have been quasi-menopausal in a sense). She was obviously decrepit and marginalized within her group, living a sad, solitary, and—it appeared—increasingly useless end of life until a strange adult male invaded the langur troop and attempted to work infanticidal mayhem. "It was Sol," writes Hrdy, "who repeatedly charged this sharp-toothed male nearly twice her weight to place herself between him and the threatened baby. When the infanticidal male seized the infant in his jaws and ran off with him, Sol pursued the attacker and wrested the wounded baby back. With danger momentarily past, and the wounded infant once again in his mother's arms, old Sol resumed her diffident attitude. That an arthritic old female would become marginalized with age is scarcely surprising. More curious was Sol's transformation from decrepit outcaste to intrepid defender."[30]

It doesn't diminish Sol's courage to point out that by defending youngsters, some of whom may be their own grandchildren, Sol and other warrior grandmothers may literally be justifying their own postreproductive existence.

Good grandmothering doesn't work for many other species, however. Recall those Gombe baboons and Serengeti lions, which enjoy a genuine—albeit brief—postreproductive life span. In both these cases, mothers are prudent, living long enough to help rear their youngest offspring, but they don't substantially improve the reproduction of their already grown daughters. (Interestingly, still-breeding elderly lionesses are a different matter: they will nurse their grandchildren, thus helping the cubs to survive.)

All things considered, however, it is premature to conclude that good grandmothering solves the riddle of human menopause. Simply as a matter of pure arithmetic, it is difficult for a woman to provide enough fitness benefits to her grandchildren to make up for what she would otherwise gain by attempting to reproduce indefinitely, even until she dies in the process. And if, for whatever reason, menopause is unavoidable, *not* selected via grandmothering, and just another sign of age-related decay such as wrinkles, gray hair, and cataracts, then elderly women would still be selected to enhance their fitness in any way possible, including the granny route. Therefore, the mere fact that grandmothers are helpful doesn't necessarily explain why they are nonreproductive. Maybe good grandmothers are simply making the best of their situation and not acting out the reason for their nonreproductive state.

Moreover, the fitness benefit available to and from grandmothers in some cases isn't very impressive, and in others it is barely an option.

Consider Arctic dwellers such as the Inuit or others among whom hunting rather than gathering is predominant and grandmothers contribute very little. Anthropologists Kim Hill and Magdalena Hurtado have conducted extensive

Female langurs chasing an infanticidal male. Photograph by Sarah Hrdy, courtesy of AnthroPhoto.

studies of the Aché people, who inhabit the tropical forests of eastern Paraguay, where they rely primarily on hunting. Among the Aché, presence or absence of grandmothers makes relatively little difference to the survival of youngsters.[31] This discouraging result doesn't disavow the role of hardworking Hadza grandmothers, nor does it argue against a likely role of grandmother-as-provider in possibly selecting for human menopause. After all, the chances are that early hominid ancestors gathered at least as much as they hunted. The finding simply emphasizes that human beings—like other animals—are eminently good tacticians: they have been selected to follow a variety of paths depending on ecological and social conditions.

A More Inclusive Hypothesis

Let's grant that menopause may well function to keep middle-aged women from breeding at a time when their personal risk is going up (higher mortality) and payoff is going down (greater danger of producing underweight, genetically compromised, or otherwise defective offspring) so that women are more fit in the evolutionary sense if they care for those offspring that are already on the ground, regardless of whether selection has generated menopause as a way of facilitating that care. What about childless women? In their case, shouldn't natural selection favor any "last gasp" attempts at motherhood? To some extent, it does.

It is a general principle of behavioral ecology that older parents are willing to invest more and take greater risks than are younger ones; after all, with less potential reproductive future ahead of them, they have less to lose if they exhaust themselves or even perish in a last-ditch effort on behalf of their litter or nestfull of young. So why does menopause occur at all among middle-aged women who have never had a child?

One explanation is that a large percentage of modern childlessness is intentional, a result (for many, a reward) of birth control; the ability to determine family size and even to decline breeding altogether is, after all, quite new to the human experience. For most of *Homo sapiens'* evolutionary history, nearly every healthy woman became a mother. Accordingly, it shouldn't be expected that menopause would be so fine-tuned as to exclude those who these days are childless, or "child free."

Another explanation speaks to a major additional wrinkle in the mystery of menopause: the fact that postreproductive women can and do make additional contributions to their own biological success, whether they have actually had offspring or not. The issue here goes beyond merely shining extra, peripheral light on the enigma of menopause. It goes to the heart of evolution and of biology itself.

As currently understood, evolution isn't really about bodies and the success of those bodies. Rather, it is about genes. After all, bodies are ephemeral, transient concatenations of flesh, doomed to perish and disappear. Genes, in contrast, are possibly immortal because they have the potential to carry on and persist into future generations, whereupon they organize a certain amount of matter into new bodies, which then pursue their lives, eating and sleeping and metabolizing and moving and mating and making babies—or not. They also compete and cooperate among themselves, eventually projecting various quantities of themselves into yet further generations.

In his poem "Heredity," Thomas Hardy speaks for genes, showing a re-markable premonitory grasp of modern evolutionary biology:

I am the family face;
Flesh perishes, I live on
Projecting trait and trace
Through time to times anon,
And leaping from place to place
Over oblivion.

The years-heired feature that can
In curve and voice and eye
Despise the human span
Of durance—that is I;
The eternal thing in man,
That heeds no call to die.[32]

Although the natural world, as people encounter it, abounds in bodies, genes are what matter from evolution's perspective. They joust with one an-other via proxy structures known as organisms, whose bodies are variously en-trusted to get the job done, that "job" being to promote the success of their constituent genes. The telling aspect of genes isn't merely that they don't die, but that they are where the selective action is. In observing organisms, though, we see only bodies, bodies everywhere—climbing trees, swimming in the sea, flying, walking, running, creeping—even though at a deeper evolutionary lev-el, all those bodies are really acting out the consequences of their genes.

Much of that organism-level action concerns all the mundane realities of daily life, attending to the quotidian necessities of keeping the blood flowing and the air moving in and out, the cells humming, the kidneys filtering, and so forth. Nor is behavior irrelevant: scratching when you itch, migrating if you must, eating and breeding when possible are important evolutionary processes insofar—and *only* insofar—as they promote the eventual success of those genes that get their bodies to scratch, migrate, eat, breed, and so on. Of these ge-netically influenced behaviors, breeding is especially notable because making babies serves no purpose other than to project genes into the future. Whenever a sexually reproducing individual has a child, every gene within each parent experiences a 50 percent chance of being thereby passed along into the next generation; that's why individuals reproduce in the first place. It may also be why, paradoxically, evolution has orchestrated things such that women (more precisely, women's genes) at a certain point in their lives make more copies of themselves by having fewer children than by having more.

The key scientific insight in this regard—applied not to menopause but to altruism more generally—came from biologist William D. Hamilton, who recognized that reproduction is only a special case of the more general phenomenon of what life is all about. After all, why reproduce at all? Go back to Flo's folly: insofar as her life was shortened by having a baby in her twilight years, the same was likely true of Flo's having any babies at all. If she were really trying (i.e., selected) to optimize the situation of her body rather than that of her genes, Flo would have remained celibate or at least nonreproductive. Looked at from the viewpoint of bodies, making babies is itself a bizarre act of self-abnegating altruism: to reproduce is to create a new body or two, which in turn make substantial demands on the creator. For evolutionary biologists, of course, this argument is a reductio ad absurdum because reproducing isn't altruism at all, but rather in a genuine sense the ultimate act of selfishness whereby genes perpetuate themselves, or actually identical copies of themselves, by packaging these copies, with a 50 percent probability in each case, in precisely those new bodies known as offspring. A gene for nonreproduction would obviously have a dim evolutionary future indeed.

This was Hamilton's great insight: recognizing that reproduction is merely a special case of the more general phenomenon whereby genes look after themselves. This insight ushered in a new era in evolutionary thought, focusing on the machinations by which genes promote themselves, using bodies as "survival machines."[33] Seemingly altruistic behavior such as food sharing, alarm calling in response to a predator, and even in some cases refraining from breeding altogether can be seen as fitting under a new explanatory umbrella—or, actually, the same old powerful Darwinian one, differential reproduction. But here we're talking about the reproduction of genes rather than of bodies.

And what of the apparent altruism of individuals who go out of their way to help others, paradoxically losing their own fitness as a result? Under a gene-focused perspective, not only does this behavior make sense, but the paradox erases itself and is replaced with a new, more inclusive view of fitness according to which natural selection, operating upon genes, generates actions that seem to be altruism from the perspective of bodies, but that are actually selfishness at the level of genes. The traditional pre-DNA, Darwinian view of fitness as simple reproductive success has therefore been eclipsed by a more sophisticated but nonetheless still Darwinian view of "inclusive fitness" that looks instead at genes.

What does all this have to do with grandmothers? Everything (or, at least, quite a bit).

For instance, inclusive fitness theory—or, as it is often called, *kin selection*—can go a long way toward showing how nonreproductive women can profit evolutionarily from menopause. Keep in mind that bearing children isn't the only way that living things achieve genetic success. Inclusive fitness is inclusive because it embraces any actions that contribute to the success of *all* identical copies of one's genes, not just those encased in children. They might be in grandchildren, cousins, nieces, nephews, and so forth, with the importance of such bodies devalued in proportion as they are more distantly related because a "closer" relative means a higher probability of shared genes. Among many animals, the epitome of "altruism" takes place when individuals do not reproduce at all, as when worker bees refrain from breeding and instead labor—celibate yet busy—for the reproductive success of the queen and her other progeny. A childless, postmenopausal woman is hardly a worker bee, but, like her insect counterpart, she can nonetheless enhance the long-term success of her genes by contributing to those same genes packaged in other, nonoffspring bodies. Hence, it isn't literal altruism—that is to say, genetic sacrifice—when she lends a hand to her nieces, nephews, cousins, siblings, and so on.

Moreover, there is no requirement that a helping elder be a maiden aunt who lacks progeny. She might also be a highly prolific Hadza matriarch, who returns from a day's hard digging with more food than her grandchildren can use and who—no surprise to evolutionary theorists—distributes the bounty to her extended family.

Jared Diamond has written convincingly about the benefits that the elderly—most often old women—convey to others. Before he became known as the author of such ecologically insightful tomes as *Collapse* and *Guns, Germs, and Steel,* Diamond had already made a name for himself among ornithologists and biogeographers for his research on the birds of New Guinea and the South Pacific. He recounts how in searching for tribal lore about the natural history of a particular animal, he was often led "to a hut, inside of which is an old man or woman, often blind with cataracts, barely able to walk, toothless, and unable to eat any food that hasn't been prechewed by someone else. But that old person is the tribe's library. Because the society traditionally lacked writing, that old person knows much more about the local environment than anyone else and is the sole source of accurate knowledge about events that happened long ago." Diamond points out that such knowledge isn't merely of sentimental value. It can have life-or-death consequences. His account is worth repeating, despite its length, because it puts flesh on the bones of an argument that, although familiar and important, often lacks empirical substance:[34]

[I]n 1976 I visited Rennell Island in the Solomon Archipelago, lying in the Southwest Pacific's cyclone belt. When I asked about consumption of fruits and seeds by birds, my Rennellese informants gave Rennell-language names for dozens of plant species, listed for each plant species all the bird and bat species that eat its fruit, and stated whether the fruit is edible for people. Those assessments of edibility were ranked in three categories: fruits that people never eat; fruits that people regularly eat; and fruits that people eat only in famine times, such as after—and here I kept hearing a Rennell term initially unfamiliar to me— after the *hungi kengi*. Those words proved to be the Rennell name for the most destructive cyclone to have hit the island in living memory—apparently around 1910, based on people's references to datable events of the European colonial administration. The hungi kengi blew down most of Rennell's forest, destroyed gardens, and drove people to the brink of starvation. Islanders survived by eating the fruits of wild plant species that normally were not eaten, but doing so required detailed knowledge about which plants were poisonous, which were not poisonous, and whether and how the poison could be removed by some technique of food preparation.

When I began pestering my middle-aged Rennellese informants with my questions about fruit edibility, I was brought into a hut. There, in the back of the hut, once my eyes become accustomed to the dim light, was the inevitable, frail, very old woman, unable to walk without support. She was the last living person with direct experience of the plants found safe and nutritious to eat after the hungi kengi, until people's gardens began producing again. The old woman explained to me that she had been a child not quite of marriageable age at the time of the hungi kengi. Since my visit to Rennell was in 1976, and since the cyclone had struck sixty-six years before, around 1910, the woman was probably in her early eighties. Her survival after the 1910 cyclone had depended on information remembered by aged survivors of the last big cyclone before the hungi kengi. Now the ability of her people to survive another cyclone would depend on her own memories, which fortunately were very detailed.[35]

In small-scale societies such as the Rennellese, most people are related to each other. The potentially life-saving wisdom of the old lady, whose name may never be known but whose experience might be responsible for saving numerous entities with her DNA genetically inscribed upon them, might be more than enough to warrant natural selection's keeping her around despite being nonreproductive. She may have lost the ability to get pregnant, but no matter: she is nevertheless pregnant with crucial, life-preserving wisdom.

There seems little doubt that, at least in traditional societies, elders have something of value to provide to others. And just as inclusive fitness theory shows how those "others" needn't be limited to one's direct descendants, the survivor of a *hungi kengi* needn't limit her provisioning to a day's harvest of tubers.

Recall the enigmatic case of short-finned pilot whales, a species that seems to experience menopause. Interestingly, this species sometimes makes headlines when dozens end up beached and at risk of imminent death. These animals often travel in large pods of close relatives, and when tides are atypical, it appears that they sometimes get lethally disoriented. Here, then, is a prediction: when pilot whale pods suffer a dangerous lack of good piloting, it is associated with a dearth of wise, experienced, and postmenopausal old cetacean salts. Or consider a population of elephants that might experience a life-threatening drought every fifty years; at such a time, wouldn't it be a good thing if someone in the herd remembered the previous time such a drought happened and knew the only water hole likely to be retaining any water? If menopause or something resembling it is eventually discovered in other animals, the most likely candidate species would be those that associate in genetically related social groups, as opposed to being solitary, and among whom wisdom counts. (East African female elephants have a maximum life span of about sixty years; around age fifty-five, an age attained by only about one-twentieth of the population, surviving females still retain fully half of their reproductive capacity, so if they experience "menopause," it's a mere shadow of the human situation.)[36]

A case can be made that in today's technological human societies, where elders currently rely on their grandchildren to program their VCRs, translate their e-mail, and unravel ever-changing iPod mysteries, maybe elders provide less benefit than they did to the inhabitants of Rennell Island. Or maybe not. It seems undeniable that wisdom, transferable across generations, has more value than mere knowledge, which in turn is potentially more precious than naked technique, disconnected, unseasoned, and lacking perspective. If so, then the social value of postreproductive individuals would be beyond dispute, along with their biological payoff. It is also questionable whether human beings *ever* become truly postreproductive, insofar as they can continue to contribute not only to their offspring, but also to others within their inclusive purview on this side of the grave.

Minimizing Competition?

Many roads lead to menopause, or at least point in that direction: among them, the most plausible are prudent mothering, grandmothering, and kin selection, each spiced with a soupçon of senescence, and none of them exclusive of the others. Even though grandmothers clearly have a positive effect on the success of their grandchildren, it appears that this benefit isn't large enough to explain why the onset of human menopause is so early. To militate against reproducing, the grandmother effect must be truly robust, but although it is real, it doesn't seem to be *that* powerful. After all, under inclusive fitness considerations, grandchildren count for only one-half that of children, and grandmothers would receive a fitness payoff from their grandchildren even if they don't become nonreproductive and devote themselves to helping. In the inclusive fitness, evolutionary calculus, the key is the *additional* payoff received as a result of a given act. Foregoing one's own reproduction is a tough act to follow—or, rather, to compensate for.

British biologists Michael Cant and Rufous Johnstone have recently proposed another compensation, one that is enough, perhaps, to put the benefit of menopause over the top.[37] They suggest that the current theories for menopause ignore something important: the possible impact of reproductive competition, whereby breeding by one individual can depress that of another, especially if resources are scarce and life is a zero-sum game, which it often is.

Their idea is essentially that menopause has been selected for in *Homo sapiens* as a way of avoiding reproductive competition between middle-aged and younger women by having the former give way to the latter. This isn't to say that there are winners and losers, however, because, according to the competition avoidance hypothesis, by foregoing ovulation, middle-aged women gain more fitness than they would by continuing to reproduce directly. Everyone "wins." (In this sense, all of the hypotheses herein considered are similar in that each depends for its plausibility on the prospect that individuals—and genes—that follow the specified path leave more descendants than if they were to do otherwise.)

Although the life spans of mothers and daughters overlap greatly, as do—to a lesser extent—those of grandmothers and granddaughters, there is very little overlap between the generations when it comes to active reproduction. And maybe that is the point if menopause is an adaptation to minimize reproductive competition. One way to evaluate Cant and Johnstone's competition avoidance hypothesis would be to compare the reproductive success of young women with and without the presence of their own reproducing mothers. The prediction is that in the former case, mother-daughter competition would reduce the younger women's child-producing effectiveness. But because menopause occurs so reliably, there are no traditional societies (and very few individual families) in which young mothers and *their* mothers commonly seek to rear offspring simultaneously. Polygyny is quite common, though, even in the twenty-first century, and the evidence is convincing that when cowives reproduce at the same time in the same household, their offspring are less successful, presumably because of the ensuing competition.[38]

But why do the older women give way to the younger, rather than vice versa, as in other cooperatively breeding species? After all, each woman is more closely related to herself than to her daughter or her mother. As inclusive fitness theory teaches, the genetic payoff that a mother derives from successful breeding by her own daughter is 0.25, whereas that of producing yet another daughter would be 0.50. (These numbers are coefficients of genetic relationships, used to calculate inclusive fitness: 1.0 implies genetic identity; 0 indicates complete unrelatedness; and values in between indicate the degree of relatedness, with higher numbers characterizing closer relatives.) Older and still vigorous senior mothers nearly everywhere are socially dominant over young adults. Not surprisingly, therefore, among all other cooperatively breeding vertebrates, it is the *older* females who breed and the younger ones who help. In people, however, it's the other way round, with the seniors relegated to alloparenthood.

Cant and Johnstone have an explanation, which, albeit complex, rewards our attention. They note that "our closest primate relatives, chimpanzees, bonobos and gorillas, are unusual among primates because they exhibit female biased dispersal, and male dispersal is rare."[39] In other words, females tend to leave their birthplace, taking up residence with their "husband" and his family. In such cases, young females begin their reproductive lives surrounded by nonrelatives, and any kin selection expert knows[40] that in such cases evolution will favor reproduction by one's self rather than solicitude for the breeding of others. Such individuals are therefore disinclined to "altruism," whether manifested directly by helping others or indirectly by restraining one's own childbearing. As time goes on, however, there is an increase in the genetic relatedness between the aging female and any offspring produced in her social group.

In the case of humans, this relatedness would mean that as a matriarch grows older, she would be increasingly inclined to defer reproduction to her daughter-in-law, who, surrounded by nonrelatives, gains nothing by refraining from reproducing, whereas the mother-in-law stands to gain when her own son produces offspring by that daughter-in-law. Even allowing for some chance of occasional adultery, it is the latter, interestingly, who "wins" the conflict over who gets to breed. Actually, as already noted, both win insofar as the matriarch is more fit if she menopausally refrains from breeding.

This idea cries out for assessment, which won't be easy. Not only are there precious few cases of generational overlap among breeding women, but it is unknown and perhaps unknowable whether early hominids were male or female biased in their dispersal (with male-biased dispersal, the inclusive fitness considerations are rather different). Other questions remain, not least why other animals—who experience as much if not more competition among juveniles—have not been selected to avoid reproductive overlap.

There are few things as intellectually appealing as explanatory clarity, being able to state with confidence that *this* and not *that* is the cause of something. But the world needn't oblige. "We should make things as simple as possible," as Einstein once put it, "but not simpler." Especially in the multifactorial realm of biological causation, different factors are likely to be responsible in differing degrees and at varying times. As a result, many roads may indeed have led to menopause, and, unlike a normal traveler, *Homo sapiens* may have journeyed down several at once, gaining momentum from taking a mix of different paths—all at the same time.

A Tale of Two Longevities

As we have just seen, when it comes to women's reproductive system versus men's, the latter version "lives" longer than the former. As for bodies, however, it's the other way round. Whatever the cause, there can be no denying that women live longer than men. In the United States, the average life span for women is seventy-nine years; for men, it is seventy-two.

Why do women outlive men?

Even though women are the "weaker sex" when it comes to lifting weights, throwing objects, or digging ditches, a cross-cultural perspective affirms that women are biologically stronger than men in the way that really counts: from Arabia to Zaire, women live longer. In Japan and northern Europe, where life spans are especially long (because of high living standards, including good nutrition, quality medical care, and proactive public-health programs), the gap between women and men is greater than in developing countries, where women are more likely to die during childbirth and as a result of botched abortions.

Male hormones may be involved in the difference. A study of mentally retarded men who had been castrated early in life (and thus deprived of most testosterone) showed that they lived nearly fourteen years longer than similar men who were intact.[41] This finding is forty years old and needs to be replicated to be believed, yet there is evidence that testosterone inhibits the immune system, which would make its presence an added, proximal burden with which evolution has saddled males.[42] (By the same token—or, more accurately, the other side of the hormonal coin—it can be argued that to some degree estrogens act to promote long life among women.)

In any event, if a direct connection between male hormones and higher mortality is genuine, there may be an indirect one as well because testosterone exerts behavioral effects even as it influences the physiological processes of growth and metabolism. Moreover, there is an evolutionary reason why the predominant male sex hormone has the effect that it does.

Hormones alone, in fact, seem unlikely to explain why boys are so much more vulnerable than girls and why during every decade of life males die at a higher rate than females, even before "raging hormones"—most of which are male—become potentially relevant.

The difference actually begins even before birth. Boys are more vulnerable to an array of prenatal problems. For example, although many more boys than girls are conceived (approximately 120 to 150 males for every 100 females), the ratio of boys to girls at birth drops to about 105 to 100.[43] In other words, males are far more likely to be spontaneously aborted than females. In the birth process itself, boys are more susceptible to injury, perhaps because they are somewhat larger and thus have a more difficult journey through the birth canal.

During infancy, the pattern persists. Boys die more often than girls, resulting in a ratio of male to female mortality of 1.27—that is, 127 baby boys die for every 100 baby girls. Throughout life and for most ailments, this sex-skewed ratio persists: more males than females die. For example, the ratio for diabetes is 1.02 (that is, 102 males die of diabetes for every 100 females); for all cancers combined, the ratio is 1.51; pneumonia and influenza, 1.77; heart disease, 1.99; cirrhosis of the liver, 2.16; accidents, 2.93; suicide, 3.33; lung cancer, 3.43; homicide, 3.86. Some of these deaths are socially induced, at least in part, but this does not mean that they aren't evolutionarily induced as well. Males smoke more, drink more, and are generally bigger risk takers, and therefore they are more prone to accidents. Once it became socially acceptable for women to smoke, rates of lung cancer among women tragically began to approach those of men.

Nevertheless, the ten most deadly diseases—from heart attacks to pneumonia and cancer—attack men more than twice as often as women. Those diseases that predominate among women are often specific to the female body plan, such as breast cancer and bladder infections, or derive from hormonal differences, such as osteoporosis. An intriguing and as yet unexplained exception involves certain diseases of the immune system such as lupus, multiple sclerosis, and myasthenia gravis, all of which are more common among women.

Men generally have higher frequencies of drug and alcohol abuse and of "conduct disorders" such as burglary, rape, torture of animals, and a range of antisocial, often quasi-criminal behaviors generally described as "sociopathy." Men are also more likely to suffer from learning disabilities, attention deficit hyperactivity disorder, autism, and pervasive developmental disorders. By contrast, women outnumber men in depression, anxiety disorders, and so-called somatization illnesses, as well as hypochondriasis. Both sexes are equally likely to suffer from manic-depressive illness, schizophrenia, and obsessive-compulsive disorder. Whereas women are more likely to suffer depression, men are more likely to commit suicide. Women "attempt" suicide more often, typically by drug overdose, and engage in so-called parasuicidal behaviors, including wrist lacerations and other self-destructive but nonlethal behaviors. A suicidal man, by contrast, is more likely to jump off a bridge or put a bullet in his brain. Why this higher "success rate" for men? Because they use more violent and irreparable means. Furthermore, conduct disorders and drug or alcohol dependence makes people at especially high risk for suicide.

When it comes to disease resistance alone, no one knows why women are so much stronger than men. One often-suggested possibility relates to the genetics of maleness versus the genetics of femaleness. Because women have a double dose of X chromosomes, they are generally protected against the action of harmful genes present on the other X. Healthier genes tend to be dominant and can mask the action of their less healthy alternative, so women's second X buffers the effects of any possible hurtful mutation. Compared to the X, the Y chromosome—the one determining maleness—is a biological slacker, almost lacking in genetic material. Indeed, some geneticists maintain that it is on the way out, perhaps like the appendix.[44] According to this argument, males, which are XY, have "unprotected" X chromosomes and are therefore more liable to be victimized by any harmful genes present on that lonely, unpaired X, whereas women are provided with an extra reservoir of backup DNA in the event of faulty genes.

Most mutations are deleterious rather than fitness enhancing. No surprise here, considering that a mutation is by definition a random error. Given that living things are highly nonrandom structures, built up over eons of evolutionary trial and success, the likelihood is that any random change in such a highly structured machine will make it run worse, not better. Consider, for example, the likely effect of removing part—even a very small part—of a computer's innards and replacing it with some metal and plastic chosen by chance alone; the computer's performance would probably *not* improve.

Most mutations are recessive, however, so their effects are generally minimal because they can be overridden by alternative, dominant alleles. But because males are stuck with that nearly useless, male-determining Y, any fitness-reducing mutation on their X is liable to express itself rather than be protectively overridden, as in females. According to the backup chromosome hypothesis, therefore, women live longer than men because, to use a sports metaphor, they have a deeper bench.

Alternatively, greater female longevity might be attributable to the protection of women not by their genome, but rather via social traditions and cultural expectations. Women might therefore live longer on average than men because they are expected to pursue less physically dangerous, exhausting, and possibly life-shortening activities. Although differing social roles and tasks undoubtedly account for some of the life span differences between women and men, that same seven-year male-female difference in life span has interestingly been found to persist among American nuns and monks who do much the same things and are generally isolated from the hurly-burly of modern life.[45] It remains possible, nonetheless, that even in a cloistered environment women live longer not because—or not *just* because—they are biologically stronger, but because they do things less likely to kill them early.

Which brings us back to evolution and sex.

As already noted, biology defines female *versus* male, with females identified as those individuals that specialize in producing large gametes, eggs, necessarily in small numbers, whereas males make small gametes, sperm, in large quantities. This is why biologists can distinguish male from female birds, for example, even though for most species, there are no external genitalia; ditto for male versus female oysters. Sex is not a matter of breasts and beards, vaginas and penises, but of eggs and sperm. Once individuals are specialized as egg makers, they are likely to specialize as well in doing a good job at nurturing and rearing those expensive gametes. Sperm makers, in contrast, are likely to specialize in gaining access to those valuable eggs—which is to say, they are more evolutionarily fit in proportion as they are able to mate with more than one female. But because other sperm makers are selected to do the same thing, males are selected to compete among themselves in two ways: being attractive to females and coming out on top in head-to-head competition with other males.

In the world of animals in general and of polygynous (harem-forming) mammals in particular, males are therefore not surprisingly the ones who huff and puff, try to blow each other down, and beat each other over the head as they seek to gain access to females by intimidating and, if necessary, physically defeating their rivals. Even though most men stop short of assault and battery, not to mention murder, the reality is that men are far more inclined than women to behave violently and to be victimized by the violence of other men.[46] And even when no violence is employed, there seems little doubt that the male behavioral style is likely to result in more frequent accidents, injuries, and earlier death than is its female equivalent. Moreover, men are also inclined—certainly more so than are women—to engage in the kind of risky, show-off behavior that is likely to attract the attention of the opposite sex and that, in the process, results in more fractures, concussions, and assorted injuries, at least some of which are likely to be life threatening or at least life shortening.[47]

Among the more competitive sex, playing their reproductive game for higher stakes (in the extreme cases, becoming either a harem master with lots of mates or the unsuccessful bachelor with none), it is not surprising that boys and men's flamboyant, eye-catching behavior also includes its more pathological extremes. To be reproductively successful, men—more than women—need to stand out from their colleagues and competitors. Maybe in their efforts to do so, some men go over the edge. Notably, this gender gap peaks at late adolescence and early adulthood, precisely when risk taking would be expected to be highest among males of any polygynous mammal. It is no coincidence, for example, that automobile insurance is most expensive for unmarried males younger than age twenty-five. And thus no coincidence that women live longer than men.

This competition hypothesis for differences in life span focuses on men rather than on women, explaining greater female longevity by noting that men are prone to dying earlier because of their show-offy, risk-taking inclinations; another way to put this is to say that women live longer because they aren't men. There is, however, a more female-centered perspective, already mentioned briefly in chapter 5: the grandmother hypothesis and its derivatives, not simply as a possible explanation for menopause, but for extended female life span. Thus, insofar as women are guaranteed genetic relatedness to their offspring and thus to their grandchildren,[48] they may have been selected for a long postreproductive life, more than men were. So perhaps women outlive men because the genetic payoff for doing so has long been greater for women than for men. Maybe elderly women can thank their grandchildren (plus evolution) for their own longevity, no less than for their nonreproductive status.

Of Brains and Childhood

Matters of menopause and of longevity speak to some of the most basic aspects of biology, collectively known as life-history strategies. When to start reproducing, when to stop (if at all), how large to grow, how often to reproduce: these questions and others, although solved by living things via natural selection, have generated impressive computer modeling and vigorous theorizing on the part of mathematically inclined ecologists, all trying to figure out what evolution has already figured out through trial and error as well as the differential success of those combinations that have worked best.

Among those strategies identified by theorists is a continuum of options that range from so-called r-selection at one end (r stands for "reproductive rate") to K-selection at the other (K—peculiarly—stands for "carrying capacity"; it should be c). Mice are r-selected compared to elephants, which is to say that they reproduce early and often, cranking out large litters, and, of necessity, not investing very much in any one of their children. By contrast, K-selected species—elephants compared to mice—take their time becoming sexually mature, produce fewer offspring per pregnancy, invest more in each one, and don't breed all that often. Unlike their r-selected counterparts, which frequently experience boom-or-bust conditions that favor the ability to make lots of kids and quickly, K-selected species tend to maintain population levels that are more constant and that favor quality over quantity.

Human beings are generally K-selected, and the fact that they have evolved menopause fits right into this strategic mold. To some extent, within the same species, males are r-creatures, and females K-creatures. Thus, sperm makers are eerily like mice in that they make huge numbers of tiny sperm, but don't take any one of them very seriously, whereas egg makers are proportionately pachydermic in that females make a much smaller number of larger "offspring" and are likely to invest heavily in each one.

In the long run, to be sure, men and women are the same species, even though they pursue somewhat different reproductive tactics. And in that same long run, all living things are selected to produce the maximum number of descendants they can manage; a quick-and-dirty assessment would therefore suggest that all animals should to some extent be r-selected insofar as r-strategists maximize their reproductive output. But elephants don't breed like rabbits, just as mice don't breed like codfish (which produce literally hundreds of thousands of eggs at a single spawning), even though all of these species are undoubtedly breeding as rapidly as they can. The goal isn't simply to generate offspring, but rather to produce *successful* offspring, and if you're a mouse, success is best achieved by living fast, loving hard, and dying young—and producing lots of offspring whom you then abandon early to take care of themselves, while you proceed to breed again and again. If you're an elephant, success requires a very different reproductive style. And if you're a human being, you are more elephantine than muscine.

In fact, if you are a human being, especially if you are a female human being, it may well be that your K-strategy is particularly well honed to the degree that you have been selected to stop reproducing altogether at a certain point in your life so as to invest all the more in the offspring you may have already produced (prudent mothering), in those offsprings' offspring (grandmothering), and, while you're at it, in anyone else with whom you are likely to share genes (kin selection). Along the way, you also perhaps minimize competition with your relatives (competition avoidance). Part of this strategy is closely tied to another K-selected life-history tactic employed by *Homo sapiens:* producing offspring that grow slowly, receive a great deal of parental (and grandparental) investment, and take a long time to become sexually mature. Once again, think elephants versus mice and ask whether grandmothers in particular might not be especially responsible for the fact that young people are as they are: big-brained, slow-growing, investment-absorbing little darlings whose compelling, demanding, intelligent, increasingly educated way of life may owe a huge amount to doting grandparents, especially postmenopausal grandmothers.

So if you like being human—which is to say, being smart—thank your grandmother.

In one of Kipling's Just-So Stories, "The Elephant's Child," we learn how the "elephant's child," overflowing with curiosity, stuck his short, stubby nose into the "great grey-green, greasy Limpopo River," only to have it bitten by a crocodile and stretched into a trunk. For those similarly endowed with biological curiosity, here is some good news: more enigmas remain to be pondered, but, compared to the tribulations of Kipling's petite pachyderm, any such excursions will be less painful and more fulfilling, stretching one's consciousness rather than one's proboscis.

Epilogue

The Lure of the Limpopo

The brief poem that accompanies Kipling's story of how the elephant got his trunk includes the following paean to curiosity:

I keep six honest serving-men
(They taught me all I knew);
Their names are What and Why and When
And How and Where and Who.[1]

Those "honest serving-men" do double duty. They not only constitute the lure of the Limpopo—a metaphor for the mysterious, seductive unknowns of nature—but also guide everyone's inner Elephant Child on his or her quest to unravel those puzzles, enigmas, and mysteries that make up the natural world.

All too often students are reluctant to ask a question unless it is a "good question," which is to say one that shows how much the questioner already knows. As a corrective, recall the old saw, "The only stupid question is one that goes unasked," and bear in mind that like many old saws, this one still has some sharp teeth. What and Why and When and

Rudyard Kipling's original drawing of "The Elephant's Child" getting its nose stretched into a trunk, thereby emphasizing the downside of succumbing to what we call "the lure of the Limpopo" (although Kipling also includes, reassuringly, a "Bi-Coloured-Python-Rock-Snake," swimming to the elephant's assistance). We maintain that scientific curiosity has its upside as well, although not one that readily lends itself to a single illustration, however semi-zoological. Image scanned by George P. Landow, http://www.victorianweb.org/art/illustration/kipling/15.html.

How and Where and Who are the kind of honest advisers and legitimate questions that prod people to propose Just-So Stories, then to struggle with possible answers—in other words, to poke around in the Limpopo River and "do" science.

This brings us to a final question. In a world beset by more than its share of tragedy, desperation, and well-founded anxiety, with wars, famines, natural disasters, the looming threat of catastrophic climate change, terrorism, nuclear weapons, epidemic disease, species extinctions, plummeting biodiversity, looming resource scarcity, environmental pollution, and a plague of religious fundamentalism, interpersonal and intergroup violence, as well as catastrophic overpopulation and economic crises, why bother at all with the evolutionary enigmas of the female body?

In its own way, this "why" question is as difficult as the other mysteries just explored—maybe more so because it doesn't lend itself to an array of Just-So Stories or testable hypotheses. Nonetheless, here are some possible answers.

We should bother with these enigmas because they are there. Because understanding them and other bodily enigmas might lead to practical consequences. Because exploring them might help sharpen one's mind and enhance one's appreciation of the power of evolution not only to shape the living world, but also to enlarge the mental muscles of those inclined to wield its explanatory energy. Because it is good, clean intellectual fun, and in this world of woes, there is much to be said for intellectual inquiry and mental playfulness, if only as a temporary distraction from more consequential matters. (Cream of mushroom soup, anyone?) Because Just-So Stories, of all sorts, deserve more respect than they have received, offering as they do insights into how science works and into the curious pleasure of admitting what is *not* known, enhanced by the thrill of acknowledging the many possibilities of what might be.

And because women's bodies are exciting, more than just erotically. They have the added spice of tickling the scientific imagination with genuine unknowns, offering mysteries to be contemplated by every woman with a body and by every man or woman with the wit to wonder about them—just so.

Notes

1. On Scientific Mysteries and Just-So Stories

1. Some of these hypotheses really *are* absurd, but worth considering nonetheless.

2. Rudyard Kipling, "How the Whale Got His Throat," in *Just-So Stories* (New York: Penguin, 2000).

3. Karl Popper famously—and for the most part successfully—argued that scientific hypotheses can never be formally proved. No amount of confirmation, he claimed, can clinch the deal. At best, they can be *disproved*, and thus an idea can be considered scientific only if it can be falsified.

4. When Topsy, a slave girl in the novel *Uncle Tom's Cabin*, is asked if she knew where she came from, she famously replies: "I s'pect I growed. Don't think nobody never made me." See Harriet Beecher Stowe, *Uncle Tom's Cabin* (New York: Bantam, 1982).

5. See Robert King, Fishing into our past, *Evolutionary Psychology* 6 (2008): 365–368.

6. Natalie Angier, *Woman: An Intimate Geography* (New York: Houghton Mifflin, 1999).

7. Friedrich Nietzsche, *The Gay Science*, trans. Walter Kaufmann (1882; repr., New York: Vintage, 1974).

8. Daniel Dennett, *Darwin's Dangerous Idea* (New York: Simon & Schuster, 1996).

2. Why Menstruate?

1. Jeffrey Eugenides, *Middlesex* (New York: Farrar, Straus and Giroux, 2002).

2. Old World monkeys are primates from Africa and Asia that lack prehensile tails and are more closely related to human beings than are the New World monkeys.

3. S. B. Eaton and S. B. Eaton III, Breast cancer in evolutionary context, in *Evolutionary Medicine*, ed. W. R. Trevathan, E. O. Smith, and J. J. McKenna (New York: Oxford University Press, 1999).

4. B. I. Strassmann, The biology of menstruation in *Homo sapiens:* Total lifetime menses, fecundity, and nonsynchrony in a natural-fertility population, *Current Anthropology* 38 (1997): 123–129.

5. See S. K. Wasser and D. P. Barash, Reproductive suppression among female mammals: Implications for biomedicine and sexual selection theory, *Quarterly Review of Biology* 58 (1983): 513–538.

6. C. Knight, *Blood Relations: Menstruation and the Origins of Culture* (New Haven, Conn.: Yale University Press, 1995).

7. Pliny the Elder (Gaius Plinius Secundus), *Natural History: A Selection* (New York: Penguin, 1991).

8. M. Profet, Menstruation as a defense against pathogens transported by sperm, *Quarterly Review of Biology* 68 (3) (1993): 335–386.

9. Quoted in C. F. Fluhmann, *Menstrual Disorders: Pathology, Diagnosis, and Treatment* (Philadelphia: W. B. Saunders, 1939).

10. J. E. Markee, Menstruation in intraocular endometrial transplants in the rhesus monkey, *Contributions to Embryology* 177 (1940): 221–308.

11. A. Lethaby, C. Augood, K. Duckitt, and C. Farquhar, Nonsteroidal anti-inflammatory drugs for heavy menstrual bleeding, *Cochrane Database System Review* 4 (October 17, 2007): CD000400.

12. M. Weinstein, T. Gorrindo, A. Riley, J. Mormino, J. Niedfeldt, B. Singer, G. Rodríguez, J. Simon, and S. Pincus, Timing of menopause and patterns of menstrual bleeding, *American Journal of Epidemiology* 158 (2003): 782–791.

13. Profet, Menstruation as a defense against pathogens.

14. B. I. Strassmann, The evolution of endometrial cycles and menstruation, *Quarterly Review of Biology* 71 (1996): 181–220.

15. Ibid.

16. Ibid.

17. C. A. Finn, The meaning of menstruation, *Human Reproduction* 9 (1994): 1202–1203.

18. C. A. Finn, Menstruation: A nonadaptive consequence of uterine evolution, *Quarterly Review of Biology* 73 (1998): 163–173.

19. At least, we thought we had originated the Competence Test Hypothesis; after writing it out for this chapter, however, we encountered an article that made a similar argument, although only in passing. See J. Clarke, The meaning of menstruation in the elimination of abnormal embryos, *Human Reproduction* 9 (1994): 1204–1207.

20. C. Fiala and K. Gemzel-Danielsson, Review of medical abortion using mifepristone in combination with a prostaglandin analogue, *Contraception* 74 (1) (2006): 66–86.

21. A. J. Wilcox, C. R. Weinberg, J. F. O'Connor, D. D. Baird, J. P. Schlatterer, R. E. Canfield, E. G. Armstrong, and B. C. Nisula, Incidence of early loss of pregnancy, *New England Journal of Medicine* 319 (1988): 189–194.

22. A. T. Hertig, J. Rock, E. C. Adams, and M. C. Menkin, Thirty-four fertilized human ova, good, bad, and indifferent, recovered from 210 women of known fertility: A study of biological wastage in early human pregnancy, *Pediatrics* 23 (1959): 202–211.

23. B. E. Rolfe, Detection of fetal wastage, *Fertility and Sterility* 5 (1982): 655–660.

24. R. L. Trivers, Parent-offspring conflict, *American Zoologist* 14 (1974): 249–264.

25. Strassmann, The biology of menstruation in *Homo sapiens*.

26. K. Yamazaki and G. K. Beauchamp, Genetic basis for MHC-dependent mate choice, *Advances in Genetics* 59 (2007): 129–145.

27. I. Savic, H. Berglund, and P. Lindström, Brain response to putative pheromones in homosexual men, *Proceedings of the National Academy of Sciences (USA)* 102 (20) (2005): 7356–7361.

28. M. K. McClintock, Menstrual synchrony and suppression, *Nature* 229 (1971): 244–245.

29. W. K. Whitten, Modification of the oestrous cycle of the mouse by external stimuli associated with the male, *Journal of Endocrinology* 13 (1956): 399–404; W. K. Whitten, Effect of exteroceptive factors on the oestrous cycle of mice, *Nature* 180 (1957): 1436.

30. W. K. Whitten, F. H. Bronson, and J. A. Greenstein, Estrus-inducing pheromone of male mice: Transport by movement of air, *Science* 161 (1968): 584–585.

31. K. Stern and M. K. McClintock, Regulation of ovulation by human pheromones, *Nature* 392 (1998): 177–179.

32. J. Surowieckim, *The Wisdom of Crowds* (New York: Anchor, 2005).

33. R. B. Zajonc, Social facilitation, *Science* 149 (1965): 269–274.

34. A. N. Whitehead, *Science and the Modern World* (New York: Simon & Schuster, 1925).

35. W. R. Trevathan, M. H. Burleson, and W. L. Gregory, No evidence for menstrual synchrony in lesbian couples, *Psychoneuroendocrinology* 18 (1993): 425–435.

36. Strassmann, The biology of menstruation in *Homo sapiens*.

37. A sample of those in favor: A. Weller and L. Weller, Menstrual synchrony between mothers and daughters and between roommates, *Physiology and Behavior* 53 (1993): 943–949; A. Weller and L. Weller, The impact of social interaction factors on menstrual synchrony in the workplace, *Psychoneuroendocrinology* 20 (1995): 21–31; A. Weller and L. Weller, Examination of menstrual synchrony among women basketball players, *Psychoneuroendocrinology* 20 (1995): 613–622; A. Weller and L. Weller, Menstrual variability and the measurement of menstrual synchrony, *Psychoneuroendocrinology* 22 (1997): 115–128; A. Weller and L. Weller, Prolonged and very intensive contact may not be conducive to menstrual synchrony, *Psychoneuroendocrinology* 23 (1998): 19–32.

A sample of those opposed: Anna Ziomkiewicz, Menstrual synchrony: Fact or artifact? *Human Nature* 17 (4) (2006): 419–432; Z. Yang and J.C. Schank, Women do not synchronize their menstrual cycles, *Human Nature* 17 (4) (2006): 434–447; J.C. Schank, Do human menstrual-cycle pheromones exist? *Human Nature* 17 (4) (2006): 448–470; B.I. Strassmann, Menstrual synchrony: Cause for doubt, *Human Reproduction* 14 (1999): 579–580; H.C. Wilson, A critical review of menstrual synchrony research, *Psychoneuroendocrinology* 17 (1992): 565–569; H.C. Wilson, S.H. Kiefhaber, and V. Gravel, Two studies of menstrual synchrony: Negative results, *Psychoneuroendocrinology* 16 (1991): 353–359.

3. Invisible Ovulation

1. John Berryman, "Dream Song 4," in *John Berryman: Selected Poems* (New York: Library of America, 2004).

2. L.L. Sievert and C.A. Dubois, Validating signals of ovulation: Do women who think they know, really know? *American Journal of Human Biology* 17 (3) (2005): 310–320.

3. Interestingly, several suggestions have been made as to the adaptive significance of human eye color: that dark irises make it more difficult to determine the degree of pupil dilation, that blue eyes see better at the red end of the visible spectrum, and so on. To our knowledge, all of these suggestions are at present additional Just-So Stories, legitimate hypotheses in need of testing.

4. In evolution-speak, there wouldn't be a good reason to lie unless liars were more successful than truth-tellers in projecting their genes into the future.

5. B. Sillén-Tulberg and A.P. Möller, The relationship between concealed ovulation and mating systems in anthropoid primates: A phylogenetic analysis, *American Naturalist* 141 (1993): 1–25.

6. M. Heistermann, K. Brauch, U. Mohle, D. Pfefferle, J. Dittami, and K. Hodges, Female ovarian cycle phase affects the timing of male sexual activity in free-ranging Barbary macaques *(Macaca sylvanus)* of Gibraltar, in *Male and Female Reproductive Strategies in Relation to Paternity Outcome in Barbary Macaques,* ed. Katrin Brauch (Göttingen, Germany: Cuvillier, 2007).

7. L.G. Domb and M. Pagel, Sexual swellings advertise female quality in wild baboons, *Nature* 410 (2001): 204–206.

8. R.D. Alexander and K.M. Noonan, Concealment of ovulation, parental care, and human social evolution, in *Evolutionary Biology and Human Social Behavior: An Anthropological Perspective,* ed. N.A. Chagnon and W. Irons (North Scituate, Mass.: Duxbury Press, 1979).

9. Once evolution's stories are based on evidence, though, they stop being Just-So Stories and start becoming regular science.

10. C. E. G. Tutin and P. R. McGinnis, Chimpanzee reproduction in the wild, in *Reproductive Biology of the Great Apes,* ed. C. E. Graham (New York: Academic Press, 1981).

11. D. Singh and P. M. Bronstad, Female body odour is a potential cue to ovulation, *Proceedings of the Royal Society of London, B* 268 (2001): 797–801.

12. S. Kuukasjärvi, C. J. Eriksson, E. Koskela, T. Mappes, K. Nissinen, and M. J. Rantal, Attractiveness of women's body odors over the menstrual cycle: The role of oral contraception and received sex, *Behavioural Ecology* 15 (2004): 579–584.

13. S. Gangestad, R. Thornhill, and C. E. Garver, Changes in women's sexual interests and their partners' mate retention tactics across the menstrual cycle: Evidence for shifting conflicts of interest, *Proceedings of the Royal Society of London, B* 269 (2002): 975–982.

14. S. C. Roberts, J. Havlicek, J. Flegr, M. Hruskova, A. C. Little, B. C. Jones, D. I. Perrett, and M. Petrie, Female facial attractiveness increases during the fertile phase of the menstrual cycle, *Proceedings of Biological Sciences Supplement* 5 (2004): S270–S272.

15. For an empirical and theoretical treatment of this phenomenon in a different species, see D. P. Barash, Mate guarding and gallivanting by male hoary marmots *(Marmota caligata), Behavioral Ecology and Sociobiology* 9 (1991): 187–193.

16. D. B. Adams, A. R. Gold, and A. D. Burt, Rise in female-initiated sexual activity at ovulation and its suppression by oral contraceptives, *New England Journal of Medicine* 299 (1978): 1145–1150; A. Brewis and M. Meyer, Demographic evidence that human ovulation is undetectable (at least in pair bonds), *Current Anthropology* 46 (2005): 465–471.

17. See, for example, S. Hrdy, *The Langurs of Abu* (Cambridge, Mass.: Harvard University Press, 1979).

18. M. Heistermann, T. Ziegler, C. P. van Schaik, K. Launhardt, P. Winkler, and J. K. Hodges, Loss of oestrus, concealed ovulation, and paternity confusion in free-ranging Hanuman langurs, *Proceedings of Biological Sciences* 268 (1484) (2001): 2445–2451.

19. D. P. Barash and J. E. Lipton, *The Myth of Monogamy: Fidelity and Infidelity in Animals and People* (New York: Holt, 2002).

20. Ibid.

21. Note that *estrus* is a noun, whereas *estrous* is an adjective.

22. R. Thornhill and C. Palmer, *The Natural History of Rape* (Cambridge, Mass.: MIT Press, 2000).

23. C. Worthman, Psychoendrocrine study of human behavior: Some interactions of steroid hormones with affect and behavior in the !Kung San, Ph.D. diss., Harvard University, 1978.

24. S. Matteo and E. F. Rissman, Increased sexual activity during the midcycle portion of the human menstrual cycle, *Hormones and Behavior* 18 (1984): 249–255.

25. E. Hampson and D. Kimura, Reciprocal effects of hormonal fluctuations on human motor and perceptual-spatial skills, *Behavioral Neuroscience* 102 (1988): 456–459; A. Bröder and N. Hohmann, Variations in risk taking behavior over the menstrual cycle: An improved replication, *Evolution and Human Behavior* 24 (2003): 391–398; D. M. T. Fessler and C. D. Navarrete, Domain-specific variation in disgust sensitivity across the menstrual cycle, *Evolution and Human Behavior* 24 (2003): 406–417; D. R. Feinberg, B. C. Jones, M. J. Law Smith, F. R. Moore, L. M. DeBruine, R. E. Cornwell, S. G. Hillier, and D. I. Perrett, Menstrual cycle, trait estrogen level, and masculinity preferences in the human voice, *Hormones and Behavior* 49 (2006): 215–222; S. W. Gangestad, J. A. Simpson, A. J. Cousins, C. E. Garver-Apgar, and P. N. Christensen, Women's preferences for male behavioral displays shift across the menstrual cycle, *Psychological Science* 15 (2004): 203–207; K. Grammer, L. Renninger, and B. Fisher, Disco clothing, female sexual motivation, and relationship status: Is she dressed to impress? *Journal of Sex Research* 41 (2004): 66–74; M. Haselton and S. W. Gangestad, Conditional expression of women's desires and men's mate guarding across the ovulatory cycle, *Hormones and Behavior* 49 (2006): 509–518; M. Haselton and G. F. Miller, Women's fertility across the cycle increases the short-term attractiveness of creative intelligence compared to wealth, *Human Nature* 17 (2006): 50–73; M. G. Haselton, M. Mortezaie, E. G. Pillsworth, A. M. Bleske-Rechek, and D. A. Frederick, Ovulation and human female ornamentation: Near ovulation, women dress to impress, *Hormones and Behavior* 51 (2007): 40–45; A. C. Little, B. C. Jones, and R. P. Burriss, Preferences for masculinity in male bodies change across the menstrual cycle, *Hormones and Behavior* 31 (2007): 633–639; A. C. Little, B. C. Jones, D. M. Burt, and D. I. Perrett, Preferences for symmetry in faces change across the menstrual cycle, *Biological Psychology* 76 (2007): 209–216.

26. G. N. Miller, J. M. Tybur, and B. D. Jordan, Ovulatory cycle effects on tip earnings by lap dancers: Economic evidence for human estrus? *Evolution and Human Behavior* 28 (2007): 375–381.

27. For example, see R. Thornhill, Human sperm competition and woman's dual sexuality, in *Sperm Competition in Humans,* ed. T. K. Shackelford and N. Pound (New York: Springer, 2006); R. Thornhill, The evolution of woman's estrus, extended sexuality, and concealed ovulation and their implications for understanding human sexuality, in *The Evolution of Mind,* ed. S. W. Gangestad and J. A. Simpson (New York: Guilford, 2007).

28. S. W. Gangestad and R. Thornhill, Human oestrus, *Proceedings of the Royal Society of London, B* 275 (2008): 991–1000.

29. Haselton and Gangestad, Conditional expression of women's desires and male mate retention efforts across the ovulatory cycle; E. G. Pillsworth and M. G. Haselton, Male sexual attractiveness predicts differential ovulatory shifts in female extra-pair attraction and male mate retention, *Evolution and Human Behavior* 27 (2006): 247–258.

30. C. E. Garver-Apgar, S. W. Gangestad, R. Thornhill, R. D. Miller, and J. J. Olp,

MHC alleles, sexual responsivity, and unfaithfulness in romantic couples, *Psychological Science* 17 (2006): 830–835.

31. A. J. Wilcox, D. D. Baird, D. B. Dunson, D. R. McConnaughey, J. S. Kesner, and C. R. Weinberg, On the frequency of intercourse around ovulation: Evidence for biological influences, *Human Reproduction* 19 (2004): 1539–1543; Brewis and Meyer, Demographic evidence that human ovulation is undetectable (at least in pair bonds).

32. M. Pagel, Evolution of conspicuous estrous advertisement in Old World monkeys, *Animal Behaviour* 47 (1994): 1333–1341.

33. Miller, Tybur, and Jordan, Ovulatory cycle effects on tip earnings by lap dancers.

34. Ibid.

35. J. Lancaster and C. Lancaster, The watershed: Change in parental investment and family formation strategies in the course of human evolution, in *Parenting Across the Human Lifespan: Biosocial Dimensions,* ed. J. Lancaster, J. Altmann, A. Rossi, and L. Sherrod (Hawthorne, N.Y.: Aldine de Gruyter, 1987).

36. S. K. Wasser and D. P. Barash, Reproductive suppression among female mammals: Implications for biomedicine and sexual selection theory, *Quarterly Review of Biology* 58 (1983): 513–538.

37. This same argument operated against the Signaling Hypothesis for menstruation.

38. However, it may be a bit of a letdown to find out that at least a few other creatures beat us to the punch in separating sex from reproduction: many bacteria, for example, regularly exchange genes when they are not dividing!

39. S. M. Platek and T. K. Shackelford, eds., *Female Infidelity and Paternal Uncertainty: Evolutionary Perspectives on Male Anti-cuckoldry Tactics* (New York: Cambridge University Press, 2006).

40. For evidence that a significant proportion of children are fathered by someone other than the mother's mate, see K. G. Anderson, How well does paternity confidence match actual paternity? Evidence from worldwide nonpaternity rates, *Current Anthropology* 47 (2006): 513–520; L. W. Simmons, R. C. Firman, G. Rhodes, and M. Peters, Human sperm competition: Testis size, sperm production, and rates of extra-pair copulations, *Animal Behavior* 68 (2004): 297–302.

41. N. Burley, The evolution of concealed ovulation, *American Naturalist* 114 (1979): 835–858.

4. Breasts and Other Curves

1. C. S. Ford and F. A. Beach, *Patterns of Sexual Behavior* (New York: Harper and Row, 1951).

2. L. Schiebinger, Why mammals are called mammals: Gender politics in eighteenth-century natural history, *American Historical Review* 98 (1993): 382–411.

3. N. Angier, *Woman: An Intimate Geography* (New York: Houghton Mifflin, 1999).

4. D. Morris, *The Naked Ape: A Zoologist's Study of the Human Animal* (New York: McGraw-Hill, 1967). Bear in mind that this statement, like other explanations that seem to imply intentionality, warrants being interpreted in evolutionary terms—for example, women whose genetic makeup resulted in enlarged breasts would somehow have been especially successful in promoting this genetic tendency. No intentional "encouragement" is implied.

5. E. Morgan, *The Aquatic Ape Hypothesis* (New York: Souvenir Press, 1997).

6. G. Becker, Nutrition for lactating women, in *Core Curriculum for Lactation Consultant Practice,* ed. M. Walker (Sudbury, Mass.: Jones and Bartlett, 2002).

7. W. James, *Psychology: Briefer Course* (1892; repr., South Bend, Ind.: University of Notre Dame Press, 1985).

8. James Joyce, *Portrait of the Artist as a Young Man* (New York: Penguin, 2005).

9. B.S. Low, Fat and deception, *Ethology and Sociobiology* 11 (1990): 67–74; B.S. Low, R.D. Alexander, and K.M. Noonan, Human hips, breasts, and buttocks: Is fat deceptive? *Ethology and Sociobiology* 8 (1987): 249–257.

10. D. Singh, Adaptive significance of female physical attractiveness: Role of waist-to-hip ratio, *Journal of Personality and Social Psychology* 65 (1993): 293–307; D. Singh, Mating strategies of young women: Role of physical attractiveness, *Journal of Sex Research* 41 (1) (2004): 43–57.

11. S.W. Gangestad and G.J. Scheyd, The evolution of human physical attractiveness, *Annual Review of Anthropology* 34 (2005): 523–548; D. Singh and P. Randall, Beauty is in the eye of the plastic surgeon: Waist–hip ratio (WHR) and women's attractiveness, *Personality and Individual Differences* 43 (2007): 329–340; V. Swami, C. Greven, and A. Furnham, More than just skin-deep? A pilot study integrating physical and non-physical factors in the perception of physical attractiveness, *Personality and Individual Differences* 42 (2007): 563–572; B.J. Dixson, A.F. Dixson, L. Baoguo, and M.J. Anderson, Studies of human physique and sexual attractiveness: Sexual preferences of men and women in China, *American Journal of Human Biology* 19 (2007): 88–95; V. Swami and M.J. Tovée, The relative contribution of profile body shape and weight to judgments of women's physical attractiveness in Britain and Malaysia, *Body Image* 4 (2007): 391–396.

12. A.M. Hurtado, K. Hawkes, K. Hill, and H. Kaplan, Female subsistence strategies among Aché hunter-gatherers of eastern Paraguay, *Human Ecology* 13 (1985): 1–28.

13. C.E. Casey and K.M. Hambidge, Nutritional aspects of human lactation, in *Lactation, Physiology, Nutrition, and Breast Feeding,* ed. M.C. Neville and M.R. Neifert (New York: Plenum, 1983).

14. C. Stanford, *Chimpanzee and Red Colobus: The Ecology of Predator and Prey* (Cambridge, Mass.: Harvard University Press, 1998).

15. B. Pawlowski and G. Jasienska, Women's body morphology and preferences for sexual partners' characteristics, *Evolution and Human Behavior* (in press).

16. S. A. LeBlanc and E. Barnes, On the adaptive significance of the female breast, *American Naturalist* 108 (1974): 577–578.

17. J. G. H. Cant, Hypothesis for the evolution of human breasts and buttocks, *American Naturalist* 117 (1981): 199–204.

18. G. Jasienska, A. Ziomkiewicz, P. T. Ellison, S. F. Lipson, and I. Thune, Large breasts and narrow waists indicate high reproductive potential in women, *Proceedings of the Royal Society of London, B* 271 (2004): 1213–1217.

19. As we'll see, however, there are some interesting variations on the theme.

20. See, for example, M. Turner, ed., *The Artful Mind: Cognitive Science and the Riddle of Human Creativity* (New York: Oxford University Press, 2006).

21. V. S. Ramachandran and W. Hirstein, The science of art: A neurological theory of aesthetic experience, *Journal of Consciousness Studies* 6 (1999).

22. H. Kawabata and S. Zeki, Neural correlates of beauty, *Neurophysiology* 91 (2004): 1699–1705.

23. V. S. Ramachandran, The emerging mind, at http://www.bbc.co.uk/radio4/reith2003/lecture3.shtml.

24. M. J. Tovée, S. M. Mason, J. L. Emery, S. E. McCluskey, and E. M. Cohen-Tovée, Supermodels: Stick insects or hourglasses? *Lancet* 350 (1997): 1474–1475.

25. D. W. Yu and G. H. Shepard, Is beauty in the eye of the beholder? *Nature* 396 (1998): 321–322; L. S. Sugiyama, Shiwiar use of waist-to-hip ratio in assessments of female mate value, *Evolution and Human Behavior* 25 (2004): 51–62.

26. A. Furnham, K. V. Petrides, and A. Constantinides, The effects of body mass index and waist-to-hip ratio on ratings of female attractiveness, fecundity, and health, *Personality and Individual Differences* 38 (2005): 1823–1834; A. Furnham, V. Swami, and K. Shah, Body weight, waist-to-hip ratio, and breast size correlates of ratings of attractiveness and health, *Personality and Individual Differences* 41 (2006): 443–454.

27. A. P. Möller and R. Thornhill, Bilateral symmetry and sexual selection: A meta-analysis, *American Naturalist* 151 (2) (1998): 174–192; G. Rhodes, S. Yoshikawa, R. Palermo, L. W. Simmons, M. Peters, K. Lee, J. Halberstadt, and J. R. Crawford, Perceived health contributes to the attractiveness of facial symmetry, averageness, and sexual dimorphism, *Perception* 36 (2007): 1244–1252; A. C. Little, C. L. Apicella, and F. W. Marlowe, Preferences for symmetry in human faces in two cultures: Data from the UK and the Hadza, an isolated group of hunter-gatherers, *Proceedings of Biological Sciences* 27 (2007): 3113–3117; G. Rhodes, The evolutionary psychology of facial beauty, *Annual Review of Psychology* 57 (2006): 199–226.

28. A. C. Little, B. C. Jones, D. M. Burt, and D. I. Perrett, Preferences for symmetry in faces change across the menstrual cycle, *Biological Psychology* 76 (2007): 209–216.

29. G. Miller, *The Mating Mind: How Sexual Choice Shaped the Evolution of Human Nature* (New York: Doubleday, 2001).

30. J. Byrd-Craven, D. C. Geary, and J. Vigil, Evolution of human mate choice, *Journal of Sex Research* 41 (1) (2004): 117–144.

31. A. P. Möller, M. Soler, and R. Thornhill, Breast asymmetry, sexual selection, and human reproductive success, *Ethology and Sociobiology* 16 (1995): 207–219.

32. F. Marlowe, The nubility hypothesis: The human breast as an honest signal of residual reproductive value, *Human Nature* 9 (1998): 263–271.

33. R. A. Fisher, *The Genetical Theory of Natural Selection* (New York: Dover, 1958).

34. M. N. Muller, M. E. Thompson, and R. W. Wrangham, Male chimps prefer mating with old females, *Current Biology* 16 (2006): 2234–2238.

35. Marlow, The nubility hypothesis.

36. D. Singh, P. Renn, and A. Singh, Did the perils of abdominal obesity affect depiction of feminine beauty in the sixteenth to eighteenth century literature? Exploring the health and beauty link, *Proceedings of the Royal Society of London, B* 274 (2007): 891–894.

37. P. J. Weatherhead and R. J. Robertson, Offspring quality and the polygyny threshold: "The sexy son hypothesis," *American Naturalist* 113 (1979): 201–208.

38. R. Brooks, Negative genetic correlation between male sexual attractiveness and survival, *Nature* 406 (2000): 67–70; M. L. Head, J. Hunt, M. D. Jennions, R. Brooks, and R. Free, The indirect benefits of mating with attractive males outweigh the direct costs, *PLoS Biology* 3 (2) (2005): e33; E. Postma, S. C. Griffith, and R. Brooks, Evolutionary genetics: Evolution of mate choice in the wild, *Nature* 444 (2006): 7121.

39. Amotz Zahavi, Avishag Zahavi, N. Ely, and M. P. Ely, *The Handicap Principle: A Missing Piece of Darwin's Puzzle* (New York: Oxford University Press, 1999).

40. A. Grafen, Biological signals as handicaps, *Journal of Theoretical Biology* 144 (4) (1990): 517–546; K. Hausken and J. Hirshleifer, The truthful signalling hypothesis: An explicit general equilibrium model, *Journal of Theoretical Biology* 228 (4) (2004): 497–511; G. Nöldeke and L. Samuelsom, Strategic choice handicaps when females seek high male net viability, *Journal of Theoretical Biology* 221 (1) (2003): 53–59; G. S. van Doorn and F. J. Weissing, Sexual conflict and the evolution of female preferences for indicators of male quality, *American Naturalist* 168 (6) (2006): 742–757.

41. James Rado and Gerome Ragni, "My Conviction," in *Hair* (RCA Victor, 1968).

42. With the possible exception of breasts, as already described.

43. S. Brownmiller, *Femininity* (New York: Linden Press, 1984); C. Travis, K. Meginnis, and K. Bardari, Beauty, sexuality, and identity: The social control of women, in *Sexuality, Society, and Feminism: Psychological Perspectives on Women,* ed. C. B. Travis and J. W. White (Washington, D.C.: American Psychological Association, 2000); and N. Wolf, *The Beauty Myth: How Images of Female Beauty Are Used Against Women* (New York: William Morrow, 2001).

44. G. Saad, Applying evolutionary psychology in understanding the representation of women in advertisements, *Psychology and Marketing* 21 (2004): 593–612.

45. J. Gottschall, Greater emphasis on female attractiveness in *Homo sapiens:* A revised solution to an old evolutionary riddle, *Evolutionary Psychology* 5 (2007): 347–357.

46. They are neither crickets (rather, katydids) nor Mormons.

47. D.T. Gwynne, Sexual difference theory: Mormon crickets show role reversal in mate choice, *Science* 213 (1981): 779–780.

48. D.P. Barash and J.E. Lipton, *The Myth of Monogamy: Fidelity and Infidelity in Animals and People* (New York: Henry Holt, 2002).

49. B.S. Low, Sexual selection and human ornamentation, in *Evolutionary Biology and Human Social Behavior: An Anthropological Perspective,* ed. N.A. Chagnon and W. Irons (North Scituate, Mass.: Duxbury Press, 1979).

50. Ibid.

5. The Enigmatic Orgasm

1. D. Symons, *The Evolution of Human Sexuality* (New York: Oxford University Press, 1979).

2. Dr. William Acton, *The Functions and Disorders of the Reproductive Organs, in Childhood, Youth, Adult Age, and Advanced Life, Considered in the Physiological, Social, and Moral Relations* (London: John Churchill, 1857).

3. N. Burley, The evolution of concealed ovulation, *American Naturalist* 114 (1979): 835–858.

4. To be distinguished from "intromittent reinforcement"!

5. A.K. Slob, M. Ernste, and J.J. van der Werff ten Bosch, Physiological changes during copulation in male and female stumptail macaques *(Macaca arctoides), Physiology and Behavior* 38 (1986): 891–895; A.K. Slob and J.J. van der Werff ten Bosch, Orgasm in nonhuman species, in *Proceedings of the First International Conference on Orgasm,* ed. P. Kothari and R. Patel (Bombay: VRP, 1991).

6. C.M. Meston, R.J. Levin, , M.L. Sipski, , E.M. Hull, and J.R. Heiman, Women's orgasm, *Annual Review of Sex Research* 15 (2004): 173–257.

7. V. Shklovsky, *Knight's Move,* trans. Richard Sheldon (Normal, Ill.: Dlakey Archive Press, 2005).

8. As we'll see, however, one related possibility is that orgasm might help to induce sex with the "right" partner.

9. G. C. Williams, *Adaptation and Natural Selection: A Critique of Some Current Evolutionary Thought* (Princeton, N.J.: Princeton University Press, 1966).

10. Symons, *The Evolution of Human Sexuality;* S.J. Gould, Freudian slip, *Natural*

History 96 (1987): 14–21; E. Lloyd, *The Case of the Female Orgasm* (Cambridge, Mass.: Harvard University Press, 2005).

11. D. P. Mindell, This land is your land, *Evolutionary Psychology* 5 (2007): 805–811.

12. Similarly, labia derive from the same embryonic tissue that in males become testicles. Does this mean that labia are disappointed, incomplete testicles, or vice versa? By the same token, maybe the penis is an incomplete clitoris.

13. K. M. Dunn, L. F. Cherkas, and T. D. Spector, Genetic influences on variation in female orgasmic function: A twin study, *Biological Letters* 1 (2005): 260–263; K. Dawood, K. M. Kirk, J. M. Bailey, P. W. Andrews, and N. G. Martin, Genetic and environmental influences on the frequency of orgasm in women, *Twin Research in Human Genetics* 8 (1) (2005): 27–33.

14. Lloyd, *The Case of the Female Orgasm.*

15. Having just warned about the danger of taking analogies as proof, we nevertheless are employing an analogy: between the diversity of human language and the diversity of orgasm. We aren't presenting the analogy as "proof," but simply as a guide to critical thinking.

16. Lloyd, *The Case of the Female Orgasm.*

17. B. Komisaruk, B. Whipple, A. Crawford, S. Grimes, W-C. Liu, A. Kalnin, and K. Kosier, Brain activation during vaginocervical self-stimulation and orgasm in women with complete spinal cord injury: fMRI evidence of mediation by the vagus nerves, *Brain Research* 1024 (2004): 77–88.

18. K. Connell, M. K. Guess, J. La Combe, A. Wang, K. Powers, G. Lazarou, and M. Mikhail, Evaluation of the role of pudendal nerve integrity in female function using noninvasive techniques, *American Journal of Obstetrics and Gynecology* 192 (5) (2005): 1712–1717.

19. K. S. Fugl-Meyer, K. Oberg, P. O. Lundberg, B. Lewin, and A. Fugl-Meyer, On orgasm, sexual techniques, and erotic perceptions in 18- to 74-year-old Swedish women, *Journal of Sexual Medicine* 3 (1) (2006): 56–68.

20. Simone de Beauvoir, *The Second Sex,* trans. and ed. H. M. Parshley (New York: Vintage, 1952).

21. S. B. Hrdy, Infanticide among animals: A review, classification, and examination of the implications for the reproductive strategies of females, *Ethology and Sociobiology* 1 (1979): 13–40.

22. S. B. Hrdy, *Mother Nature* (New York: Ballantine Books, 2000).

23. B. R. Komisaruk, C. Beyer-Flores, and B. Whipple, *The Science of Orgasm* (Baltimore: Johns Hopkins University Press, 2006).

24. C. A. Fox, H. S. Wolff, and J. A. Baker, Measurement of intra-vaginal and intra-uterine pressures during human coitus by radio-telemetry, *Journal of Reproduction and Fertility* 22 (1970): 243–251.

25. R. J. Levin, The physiology of sexual arousal in the human female: A recreational and procreational synthesis, *Archives of Sexual Behavior* 31 (2002): 405–411.

26. R. R. Baker and M. A. Bellis, Human sperm competition: Ejaculate manipulation by females and a function for the female orgasm, *Animal Behaviour* 46 (1993): 887–909.

27. Lloyd, *The Case of the Female Orgasm.*

28. D. A. Puts and K. Dawood, The evolution of female orgasm: Adaptation or by-product? *Twin Research* 9 (2006): 467–472; ellipses indicate literature citations, which have been removed for greater readability; anyone wanting to pursue the literature cited should refer to the original article.

29. R. J. Levin, Sexual arousal: Its physiological roles in human reproduction, *Annual Review of Sex Research* 16 (2005): 154–189.

30. Komisaruk, Beyer-Flores, and Whipple, *The Science of Orgasm.*

31. C. Sue Carter, L. Ahnert, K. E. Grossmann, S. B. Hrdy, M. E. Lamb, S. W. Porges, and N. Sachser, *Attachment and Bonding: A New Synthesis* (Cambridge, Mass.: MIT Press, 2006).

32. John Donne, "Funeral Elegies: On the Death of Mistress Drury," in *John Donne's Poetry* (New York: W. W. Norton, 2006).

33. A. Nin, *The Diary of Anaïs Nin*, vol. 2 (New York: Harcourt, Brace, Jovanovich, 1970).

34. D. P. Barash, *The Whisperings Within: Evolution and the Origins of Human Nature* (New York: Harper and Row, 1979).

35. A. Troisi and M. Carosi, Female orgasm rate increases with male dominance in Japanese macaques, *Animal Behaviour* 56 (1998): 1261–1266.

36. L. Eschler, The physiology of the female orgasm as a proximate mechanism, *Sexualities, Evolution, and Gender* 6 (2004): 171–194.

37. A. K. Kiefer, D. T. Sanchez, C. J. Kalinka, and O. Ybarra, Sex roles: How women's nonconscious association of sex with submission relates to their subjective sexual arousability and ability to reach orgasm, *Sex Roles* 55 (2006): 83–94.

38. L. W. Simmons, R. C. Firman, G. Rhodes, and M. Peters, Human sperm competition: Testis size, sperm production, and rates of extrapair copulations, *Animal Behaviour* 68 (2004): 297–302.

39. E. Peterson and T. Jarvi, "False orgasm" in female brown trout: Trick or treat? *Animal Behaviour* 61 (2001): 497–501.

40. R. Thornhill, S. W. Gangestad, and R. Comer, Human female orgasm and male fluctuating asymmetry, *Animal Behaviour* 50 (1995): 1601–1615.

41. R. L. Smith, Human sperm competition, in *Sperm Competition and the Evolution of Animal Mating Systems*, ed. R. L. Smith (London: Academic Press, 1984).

42. D. P. Barash, *The Survival Game: How Game Theory Explains the Biology of Cooperation and Competition* (New York: Times/Holt, 2003).

43. Thornhill, Gangestad, and Comer, Human female orgasm and male fluctuating asymmetry.

6. THE MENOPAUSE MYSTERY

6. The Menopause Mystery

1. C. J. Moss, The demography of an African elephant *(Loxodonta africana)* population in Amboseli, Kenya, *Journal of Zoology* 255 (2001): 145–156.

2. S. A. Mizroch, Analyses of some biological parameters in the Antarctic fin whale, *Report of the International Whaling Commission* 31 (1981): 425–434.

3. H. Marsh and T. Kasuya, Evidence for reproductive senescence in female cetaceans, *Report of the International Whaling Commission* 8 (1986): 57–74.

4. There is a recent report of menopause in captive gorillas; whether it occurs among free-living gorillas, however, is an open question. See S. Atsalis, S. W. Margulis, A. Bellem, and N. Wielebnowski, Sexual behavior and hormonal estrus cycles in captive aged lowland gorillas *(Gorilla gorilla), American Journal of Primatology* 62 (2) (2004): 123–132.

5. Oliver Wendell Holmes, "The Deacon's Masterpiece, or the Wonderful One-Hoss Shay," in *Illustrated Poems of Oliver Wendell Holmes* (Ann Arbor: University of Michigan Press, 2006).

6. F. Marlowe, The Patriarch Hypothesis: An alternative explanation of menopause, *Human Nature* 11 (2000): 27–42.

7. From the Jane Goodall Institute Web site at http://www.janegoodall.org/chimp_central/chimpanzees/f_family/flo.asp.

8. S. B. Hrdy, *Mother Nature: Maternal Instincts and How They Shape the Human Species* (New York: Ballantine, 1999).

9. J. Diamond, *Why is Sex Fun? The Evolution of Human Sexuality* (New York: Basic, 1997).

10. R. Sear and R. Mace, Who keeps children alive? A review of the effects of kin on child survival, *Evolution and Human Behavior* 29 (2008): 1–18.

11. J. S. Peccei, Menopause: Adaptation or epiphenomenon? *Evolutionary Anthropology* 10 (2001): 43–57.

12. Kenny Rogers, "The Gambler," on *The Gambler* (United Artists, 1978).

13. C. Packer, M. Tatar, and A. Collins, Reproductive cessation in female mammals, *Nature* 392 (1998): 807–811.

14. B. M. F. Galdikas and J. W. Wood, Birth spacing patterns in humans and apes, *American Journal of Physical Anthropology* 83 (1990): 185–191.

15. K. Hawkes, J. F. O'Connell, and N. G. Blurton Jones, Hadza women's time allocation, offspring provisioning, and the evolution of long postmenopausal life spans, *Current Anthropology* 38 (1997): 551–578.

16. S. B. Hrdy, Cooperative breeders with an ace in the hole, in *Grandmotherhood: The Evolutionary Significance of the Second Half of Female Life,* ed. E. Voland, A. Chasiotis and W. Schiefenhoevel (New Brunswick, N.J.: Rutgers University Press, 2005).

17. A. F. Russell, P. N. M. Brotherton, G. M. McIlrath, L. L. Sharpe, and T. H. Clutton-Brock, Breeding success in cooperative meerkats: Effects of helper number and maternal state, *Behavioral Ecology* 14 (2003): 486–492.

18. There is one report of male lactation: in the Dayak fruit bat, a resident of Malaysia. It is unclear, however, whether this unique case is normal or a result of unusually high phytoestrogens in the bats' environment. See C. M. Francis, E. L. P. Anthony, J. A. Brunton, and T. Kunz, Lactation in male fruit bats, *Nature* 367 (1994): 691–692.

19. M. Lahdenpera, A. F. Russell, and V. Lummaa, Selection for long lifespan in men: Benefits of grandfathering? *Proceedings of the Royal Society of London, B* 274 (2007): 2437–2444.

20. R. J. Quinlan and M. V. Flinn, Kinship, sex, and fitness in a Caribbean community, *Human Nature* 16 (2005): 32–57.

21. M. Lahdenpera, V. Lummaa, S. Helle, M. Tremblay, and A. F. Russell, Fitness benefits of prolonged post-reproductive lifespan in women, *Nature* 428 (2007): 178–181.

22. W. T. DeKay, Grandparental investment and the uncertainty of kinship, paper presented to the seventh annual meeting of the Human Behavior and Evolution Society, Santa Barbara, Calif., 1995.

23. H. A. Euler and B. Weitzel, Discriminative grandparental solicitude as a reproductive strategy, *Human Nature* 7 (1996): 39–59.

24. D. Biello, Insights: The trouble with men, *Scientific American* 297 (2007): 194–197.

25. A. Pashos, Does paternal uncertainty explain discriminative grandparental solicitude? A cross-cultural study in Greece and Germany, *Evolution and Human Behavior* 21 (2000): 97–109.

26. C. S. Jamison, L. L. Cornell, P. L. Jamison, and H. Nakazato, Are all grandmothers equal? A review and a preliminary test of the "grandmother hypothesis" in Tokugawa, Japan, *American Journal of Physical Anthropology* 119 (1) (2002): 67–76.

27. R. Sear, R. Mace, and I. A. McGregor, Maternal grandmothers improve the nutritional status and survival of children in rural Gambia, *Proceedings of the Royal Society of London, B* 267 (2000): 1641–1647.

28. L. A. Fairbanks, Vervet monkey grandmothers: Interactions with infant grandoffspring, *International Journal of Primatology* 9 (1988): 425–441; L. A. Fairbanks, Reciprocal benefits of allomothering for female vervet monkeys, *Animal Behaviour* 40 (1990): 553–562.

29. C. Packer, M. Tatar, and A. Collins, Reproductive cessation in female mammals, *Nature* 392 (1998): 807–811.

30. Hrdy, *Mother Nature*.

31. K. Hill and A. M. Hurtado, *Aché Life History: The Ecology and Demography of a Foraging People* (New York: Aldine de Gruyter, 1996).

32. Thomas Hardy, "Heredity," in *Thomas Hardy: The Complete Poems* (New York: Palgrave Macmillan, 2002).

33. R. Dawkins, *The Selfish Gene* (New York: Oxford University Press, 1989).

34. In this sense, such an argument tends to resemble the wisdom of elders more generally!

35. Diamond, *Why is Sex Fun?*

36. H. Croze, A. K. K. Hillmman, and E. M. Lang, Elephants and their habitats: How do they tolerate each other? in *Dynamics of Large Mammal Populations,* ed. C. W. Fowler and T. D. Smith (New York: Wiley, 1981).

37. M. A. Cant and R. A. Johnstone, Reproductive skew and the evolution of menopause, in *Reproductive Skew in Vertebrates,* ed. R. Hagar and C. B. Jones (Cambridge, U.K.: Cambridge University Press, 2008).

38. F. K. Amey, Polygyny and child survival in sub-Saharan Africa, *Social Biology* 49 (2002): 74–89; B. I. Strassmann, Polygyny as a risk factor for child mortality among the Dogon, *Current Anthropology* 38 (1997): 688–695.

39. Cant and Johnstone, Reproductive skew and the evolution of menopause.

40. Thanks to evolution, everyone is a kin-selection expert, regardless of whether he or she knows it.

41. J. B. Hamilton and G. E. Mestler, Mortality and survival: Comparison of eunuchs with intact men and women in a mentally retarded population, *Journal of Gerontology* 24 (1969): 395–411.

42. S. Braude, Z. Tang-Martinez, and G. T. Taylor, Stress, testosterone, and the immunoredistribution hypothesis, *Behavior and Ecology* 10 (3) (1999): 345–350.

43. J. M. Tanner, *Fetus Into Man: Physical Growth from Conception to Maturity* (Cambridge, Mass.: Harvard University Press, 1978).

44. S. Jones, *Y: The Descent of Men* (Boston: Houghton Mifflin, 2003).

45. D. L. Dingard, The sex differential in morbidity, mortality, and lifestyle, *Annual Review of Public Health* 5 (1984): 433–458.

46. M. Daly and M. Wilson, *Homicide* (New York: Aldine, 1988).

47. We recall a cartoon that appeared many years ago in *Mad* magazine. In successive frames, a bicycle-riding child exclaims, "Look, Ma, no hands!" "Look, Ma, no feet!" and then, after crashing, "Look, Ma, no teeth!" Not surprisingly, the show-off is a boy.

48. Especially grandchildren via their daughters.

Epilogue: The Lure of the Limpopo

1. Rudyard Kipling, "How the Elephant Got Its Trunk," in *Just-So Stories* (New York: Penguin, 2000).

Index